T3-BNU-288

Michael Hanke

Credit Risk, Capital Structure, and the Pricing of Equity Options

SpringerWienNewYork

Univ.-Doz. Mag. Dr. Michael Hanke
Abteilung für Quantitative Betriebswirtschaftslehre und Operations Research
Institut für Unternehmensführung
Wirtschaftsuniversität Wien, Vienna, Austria

Printing supported by the Fonds zur Förderung der wissenschaftlichen Forschung, Vienna, Austria.

Typesetting: Camera ready by author
Printing: Ferdinand Berger & Söhne Gesellschaft m.b.H., A-3580 Horn
Printed on acid-free and chlorine-free bleached paper
SPIN 10913987

With 50 Figures

ISBN 3-211-00520-X Springer-Verlag Wien New York

Preface

Beginning with the seminal papers by Black and Scholes (1973) and Merton (1973), most of the existing literature follows a standard approach for the valuation of options. This approach starts with an exogenously specified (or postulated) equity price process. Then, using hedging arguments, the fair value of an option can be derived either via partial differential equations or martingale methods, ideally as a closed-form pricing formula.

As a related field of application, Black and Scholes (1973) noted that corporate securities can be viewed (and valued) as options when the value process of the company's assets is exogenously specified and used as the underlying. This idea was taken further in work by (among others) Merton (1974), Black and Cox (1976), Geske (1977, 1979), Ingersoll (1977a,b), Galai and Schneller (1978) and, more recently, Leland (1994), Longstaff and Schwartz (1995), Leland and Toft (1996) and Ericsson and Reneby (1998).

However, as noted by Toft and Prucyk (1997, p. 1151),

> *"the direct link between financing decisions at the firm level and the pricing of derivative securities has, with few exceptions, been ignored in the theoretical and empirical option pricing literature".*

This direct link between (changes in) a company's capital structure and the pricing of options on the company's stocks will be explored in this book. In this regard, the contribution to the literature can be summarized as follows:

- It provides – for the first time – a comprehensive overview of firm value based pricing models restated in terms of modern option pricing theory. In this way, it makes the literature more accessible and yields further insights into these classical models.

- This is the first book directly addressing the link between (changes in) the company's capital structure and the prices of options on the company's stocks. Related work focuses almost exclusively on capital structure (or "structural credit risk") effects on credit derivatives.

- We show that capital structure effects potentially explain asymmetric volatility smiles, a well-known "stylized fact" in options markets.

- We provide a description of capital structure effects in option prices that may be used as a basis for option trading strategies aimed at exploiting arbitrage opportunities that cannot be detected using "traditional" option pricing models.

The book is organized as follows: Chapter 1 provides a brief review of option pricing theory under the assumption of an exogenously given stock price process. The main purpose of this chapter is an introduction to the notation used here. Some important concepts are introduced in passing; in particular, the martingale pricing approach and change of numeraire techniques. Chapter 2 motivates the subsequent analysis by extending the binomial model and the Black–Scholes model to allow for an endogenous stock price process, depending on an exogenously given process for the firm value. Chapter 3 provides a discussion of some exotic options needed as basic building blocks in later chapters, with a focus on barrier option pricing. In Chapter 4, we describe in detail a pricing framework originally presented by Ericsson and Reneby. Their probabilistic approach is the most natural way to model dependencies between a company's capital structure and the pricing of corporate securities. By deriving closed-form pricing formulae for a range of additional claims, we extend their framework considerably. In Chapter 5, we use this extended framework to review a number of classical firm value based security pricing models (or "structural credit risk models"). This alternative approach simplifies insight into the derivation of these models as well as economic interpretation. Chapter 6 presents an extension to the original framework due to Ericsson and Reneby (1996), which allows for the pricing of options on corporate securities, again using firm value as the underlying. By extending their work on firm value based option pricing, we are able not only to nest all existing firm value based option pricing models (which will be shown in Chapter 7), but also to provide an easy-to-use framework for the modelling of more complicated (and more realistic) capital structures. In Chapter 8, we apply our results to several classical capital structure models and thereby provide closed-form pricing formulae for options on equity within these well-known models. In Chapter 9, we examine static effects of firm value based option pricing. In particular, we will look at pricing biases relative to the Black–Scholes model. Chapter 10 extends this analysis to effects of changes in a firm's capital structure on the prices of equity options. A discussion of the strengths and weaknesses of the framework described here together with directions for future research concludes.

The book (my habilitation thesis) is primarily written for academics working in the fields of Finance and Mathematical Finance. It should also be accessible to the quantitatively inclined practitioner. To lower the "cost of entry" for the reader with little probabilistic background, the first chapters are devoted to introduce the necessary terminology and concepts using well-known, simple models.

Acknowledgements

A supportive work environment is essential to complete a project like this. In Vienna, where the early parts of the work were done, this was provided by the people at the Department of Operations Research, Vienna University of Economics and Business Administration. A major portion of the research has been conducted during my visit at the School of Mathematics, University of New South Wales, Sydney. I would like to thank John Price for encouraging me to apply for this position, and the faculty there for their support. Generous financial support through the *OeNB–WU–Förderungspreis*, which made this visit possible, and financial support through the *Austrian Science Fund (FWF)* for the publication of this book are also gratefully acknowledged.

Some of the research questions discussed here have come up during earlier joint work with Klaus Pötzelberger. Spiro Penev has been very helpful on some statistical problems. Many thanks to Manfred Frühwirth, Alois Geyer, Michaela Nettekoven, and Klaus Pötzelberger for giving good advice and pointing out minor errors in the final stages of preparing the manuscript. In addition, I would like to thank an anonymous referee (for the *Austrian Science Fund*) who made a number of helpful comments. Special thanks go to my wife, Konstanze, for her understanding and support during the "hot phases" of the work, and for proof-reading the manuscript. The responsibility for any remaining errors, however, remains with the author.

Contents

List of Symbols

A	event
a	number/fraction of claims
B_t	time t value of a *riskless* bond
b	number/fraction of claims
C	(discrete) coupon payment / (continuous) coupon rate
$C(\cdot)$	European call option
$\mathcal{C}(\cdot\|A)$	European call option, paying off conditional on an event A
c	number/fraction of claims
$.^{c}$	convertible security
$.^{cc}$	callable convertible security
D	face value (principal) of a bond issued by a company
$D(t)$	bond principal outstanding at time t
$D_t(\cdot)$	time t value of a bond issued by a company (which, in general, may be subject to credit risk)
d	down state
$E_t(\cdot)$	value of equity at time t
$\mathbb{E}$	expected value under (the physical) probability measure $\mathbb{P}$
$\mathbb{E}^m$	expected value under martingale measure $\mathbb{Q}^m$
$(\mathcal{F}_t)$	filtration
$G_{lI}(\cdot)$	unit down-and-in claim
$\mathcal{G}_{lI}(\cdot\|A)$	unit down-and-in claim, paying off conditional on an event A
$H(\cdot)$	European heaviside (or binary) option
$\mathcal{H}(S,K,T\|A)$	European heaviside (or binary) option, paying off conditional on an event A
$\cdot_{\cdot I}$	"in-claim"
$I_{\{A\}}$	indicator function for event A
K	strike price
L	(constant) barrier
l_t	time t value of the (time–varying) barrier
$\cdot_{l\cdot}$	"down-claim"

$\mathcal{M}$	market model
m	number of shares issued upon conversion of a convertible security
$N(\cdot)$	cumulative standard normal distribution function
$N(\mu, \sigma)$	normal distribution with mean μ and standard deviation σ
$N(a, b, \rho)$	c.d.f. of the bivariate standard normal distribution function
n	number of common stocks outstanding before conversion of a convertible security
$O(\cdot)$	asset stream
$\mathcal{O}(\cdot\|A)$	asset stream, conditional on an event A
$\cdot_{\cdot o}$	"out-claim"
$P(\cdot)$	European put option
$\mathbb{P}$	real world ("physical") probability measure
$\mathbb{P}\{A\}$	probability of event A under measure $\mathbb{P}$
p	probability of a price rise in the binomial model under $\mathbb{P}$
$\mathbb{Q}^m$	martingale measure (with security m as numeraire)
$\mathbb{Q}^m\{A\}$	probability of event A under the martingale measure $\mathbb{Q}^m$
q	probability of a price rise in the binomial model under $\mathbb{Q}$
r	risk-free interest rate
S_t	stock price at time t
$U(\cdot)$	unit stream
$\mathcal{U}(\cdot\|A)$	unit stream, paying off conditional on an event A
u	up state
$\cdot_{u\cdot}$	"up-claim"
$V_t(\phi)$	time t value of portfolio ϕ
V_t	firm value at time t
ν_t	liquidation value of the firm's assets at time t
W_t	warrant value at time t
(W_t)	Wiener process under measure $\mathbb{P}$
(W_t^m)	Wiener process under measure $\mathbb{Q}^m$
(Z_t)	stochastic process
α_t	portfolio weight (no. of stocks at time t)
β_t	portfolio weight (no. of bonds at time t)
β	rate at which free cash flow is generated by a company
$\Gamma(\cdot)$	payoff function for contingent claims which pay off at one point in time, payoff *rate* function for contingent claims which pay off as (continuous) streams

γ	dilution factor
Δ	hedge ratio
ζ	corporate tax rate
λ	market price of risk
μ	drift of a stochastic process
μ_X^m	drift of process (X_t) under measure $\mathbb{Q}^m$
ν	growth rate (for exponentially in- or decreasing claims)
$\pi_0(X)$	manufacturing costs of a contingent claim X
ρ	growth rate of an exponential barrier
σ	diffusion constant of a stochastic process
τ	first passage time of a process to a barrier
$\phi(\cdot)$	portfolio (with components in brackets)
φ^{CS}	fraction of asset value paid out to holders of corporate security CS in case of default
Ψ	contingent claim
Ω	probability space
$\Omega(\cdot)$	asset claim
ω_i	states of nature

Chapter 1

Option Pricing with an Exogenous Stock Price Process

In this chapter, we give a brief review of three standard option pricing models. We start with a very simple one-period model developed by Sharpe (1978) and Rendleman and Bartter (1979). Then, the binomial model due to Cox, Ross, and Rubinstein (1979) is briefly discussed. We conclude the chapter with an overview of the famous work by Black and Scholes (1973) and Merton (1973).

The models described in this chapter are very well-known in the finance world. The one-period and the binomial models are especially suitable (and often used in the finance literature, not only in textbooks) for the introduction of new concepts. Within these models, new ideas can be presented without running the risk of distracting the reader by other details. Here, these models serve the same purpose: Whereas in this chapter their standard versions are presented, they will be extended in Chapter 2 to introduce some of the ideas that are central to this book.

Another important purpose of this chapter is to introduce the reader to our notation. Several important concepts which will be used frequently in later chapters are also reviewed, in particular martingale (or risk-neutral) pricing and change of numeraire techniques.

1.1 A One-Period Pricing Model

The purpose of this section is threefold: First, it provides a reference model that is as simple as possible (without becoming trivial) which will be extended in Section 2.1 to present the idea of an endogenous stock price process. Second, some basic terminology and notation will be introduced.

Third, the concept of risk-neutral pricing can be conveniently presented in this setting. The following exposition draws on standard introductory texts in option pricing, see e.g. Neftci (2000), Musiela and Rutkowski (1997), Baxter and Rennie (1996), or Björk (1998).

1.1.1 Model Description

In this section, we work within a two-state, one-period, two-security market model defined on a probability space $(\Omega, (\mathcal{F}_t), \mathbb{P})$. The two possible states are denoted by ω^u and ω^d (the up- and downstate, respectively), and the associated filtration by $(\mathcal{F}_t)$.

Definition 1.1. A (one-dimensional) *stochastic process* is a mapping $S : \Omega \times \mathcal{T} \to \mathbb{R}$ such that for each fixed $t \in \mathcal{T}$, the mapping

$$S_t : \omega \mapsto S(\omega, t) = S_t(\omega) : \omega \to \mathbb{R}$$

is measurable.

In this definition, $\mathcal{T}$ denotes either an interval on the real line (for continuous-time processes; usually we take $\mathcal{T} = [0, \infty)$ or $\mathcal{T} = [0, T]$) or a set of time points (for discrete-time processes). In this section, we use discrete-time stochastic processes defined on $\mathcal{T} = \{0, T\}$. We assume that there exists a probability measure $\mathbb{P}$ on $(\Omega, \mathcal{F}_T)$ such that $\mathbb{P}\{\omega^u\}$ and $\mathbb{P}\{\omega^d\}$ are strictly positive numbers.

Definition 1.2 (Björk 1998, p. 29f., notation adapted). The symbol $\mathcal{F}_t^S$ ($\mathcal{F}_t$ for short) denotes "the information generated by (S_t) on the interval $[0, t]$ [...]". If, based upon observations of the trajectory $\{(S_s), 0 \leq s \leq t\}$, it is possible to decide whether a given event A has occurred or not, then we write this as $A \in \mathcal{F}_t$. If the value of a given stochastic variable Z can be completely determined given observations of the trajectory $\{(S_s), 0 \leq s \leq t\}$, then we also write $Z \in \mathcal{F}_t$. If (S_t) is a stochastic process such that we have $S_t \in \mathcal{F}_t \quad \forall t \geq 0$, then we say that the process (S_t) is *adapted to the filtration* $(\mathcal{F}_t)_{t \geq 0}$.

The first security (the "stock") is a strictly positive $(\mathcal{F}_t)$-adapted discrete-time stochastic process $S = (S_t)_{t \in \{0,T\}}$. Thus, $S_0 = \underline{S}$ (i.e., the starting value is deterministic), and

$$S_T(\omega) = \begin{cases} S^u & \text{if } \omega = \omega^u, \\ S^d & \text{if } \omega = \omega^d, \end{cases}$$

where $S^u > S^d$ (w.l.o.g.). The second security is a riskless bond with price process $B_0 = 1$, $B_T = 1 + r$ for some $r \in \mathbb{R}_+$. Our ultimate goal is to value derivatives or "contingent claims", as they are sometimes called.

Definition 1.3. A *contingent claim* Ψ with expiration time T is an $\mathcal{F}_T$-measurable random variable.

1.1.2 Replicating Portfolios

Let Φ denote the class of all portfolios ϕ composed of stocks and bonds with $\phi = \phi_0 = (\alpha_0, \beta_0) \in \mathbb{R}^2$. The time t value of such a portfolio will be denoted by[1] $V_t(\phi) = \alpha_t S_t + \beta_t$. The security market model $\mathcal{M}$ is completely described by the security price processes and admissible[2] portfolios ($\mathcal{M}(S, B, \Phi)$).

Definition 1.4. A portfolio ϕ is called a *replicating portfolio* for the contingent claim Ψ iff $V_T(\phi) = \Psi_T$.

Definition 1.5. A contingent claim is called *attainable* in the market model $\mathcal{M}$ iff there exists a replicating portfolio in $\mathcal{M}$.

Thus, using "natural" notation, the replicating portfolio for any contingent claim in our simple market model is given by the following linear system of equations:

$$\begin{cases} \alpha_0 S^u + B_T \beta_0 = \Psi^u \\ \alpha_0 S^d + B_T \beta_0 = \Psi^d \end{cases},$$

with unique solution

$$\alpha_0 = \frac{\Psi^u - \Psi^d}{S^u - S^d}, \quad \beta_0 = \frac{\Psi^d S^u - \Psi^u S^d}{B_T(S^u - S^d)}.$$

Definition 1.6. The *manufacturing cost* of a contingent claim Ψ in $\mathcal{M}$, $\pi_0(\Psi)$, is the initial investment needed to construct the replicating portfolio.

Thus,

$$\pi_0(\Psi) = V_0(\phi) = \alpha_0 S_0 + \beta_0 = \frac{\Psi^u - \Psi^d}{S^u - S^d} S_0 + \frac{\Psi^d S^u - \Psi^u S^d}{B_T(S^u - S^d)}. \tag{1.1}$$

For a discussion of optimality of replication, see Musiela and Rutkowski (1997, pp. 15ff.).

[1] Note that β_t denotes the *time* t *value* of the bond investment (not the *number* of the bonds!).

[2] *Admissible* refers to possible restrictions, e.g., on short selling of securities.

1.1.3 Absence of Arbitrage

Definition 1.7. A portfolio ϕ is called an *arbitrage opportunity* if $V_0(\phi) = 0$, $V_T(\phi) \geq 0$ and $\mathbb{P}\{V_T(\phi) > 0\} > 0$.

Definition 1.8. A security market model $\mathcal{M}$ is called *arbitrage-free* if it does not admit arbitrage opportunities.

In order to construct a sensible pricing model, arbitrage opportunities have to be ruled out. In particular, it should not be possible in our market model to sell any contingent claim at a price that exceeds its manufacturing cost. These restrictions are commonly known as *no-arbitrage conditions.*

Remark 1.1. It is important to distinguish between the notions of arbitrage-free *markets* vs. arbitrage-free *market models.* Whereas most people would agree that there are arbitrage opportunities in real-world markets from time to time, any model that admits such arbitrage opportunities is certainly not consistent with any meaningful notion of market equilibrium (since non-satiated investors would strive to exploit arbitrage opportunities without limits). On the contrary, a good (arbitrage-free) model should assist in *detecting* arbitrage opportunities in real-world markets.

Denote by (Ψ_t) the price process of the (attainable) contingent claim Ψ (with $\Psi_0 \in \mathbb{R}$ and $\Psi_T = \Gamma(\Psi)$, where $\Gamma(\cdot)$ denotes the *payoff function* of the contingent claim), and by Φ_Ψ the class of all portfolios $\phi_\Psi(\alpha_0, \beta_0, \delta_0)$ in stocks, bonds, and the contingent claim. Under the absence of arbitrage in our one-period market model, the manufacturing cost $\pi_0(\Psi)$ is the unique rational price of the contingent claim Ψ:

Proposition 1.1. *Given the arbitrage-free market model* $\mathcal{M}(S, B, \Phi)$, *the extended market model* $\mathcal{M}(S, B, \Psi, \Phi_\Psi)$ *is arbitrage-free iff* $\Psi_0 = \pi_0(\Psi)$.

Proof. Suppose, on the contrary, that $\Psi_0 < \pi_0(\Psi)$. In this case, an arbitrage opportunity is created by buying the contingent claim and selling the replicating portfolio given by equation (1.1). If, however, $\Psi_0 > \pi_0(\Psi)$, an arbitrage opportunity is created by selling one call option and buying the replicating portfolio given by equation (1.1). In both cases, the riskless profit is given by $|\Psi_0 - \pi_0(\Psi)|$. □

1.1.4 Risk-Neutral Valuation and Equivalent Martingale Measures

One approach to contingent claims pricing, the *martingale method* or *risk-neutral valuation method*, relies on the use of so-called *martingale measures.*

Definition 1.9. A stochastic process (Z_t) is called an $(\mathcal{F}_t)$-martingale under the probability measure $\mathbb{P}$ if the following conditions hold.

- (Z_t) is adapted to the filtration $(\mathcal{F}_t)$.
- $\mathbb{E}_\mathbb{P}[|Z_t|] < \infty \quad \forall t.$
- $\mathbb{E}_\mathbb{P}[Z_t|\mathcal{F}_s] = Z_s \quad \forall s, t$ with $s \leq t$.

Denote by $\tilde{S}$ the *discounted* price process of security S (defined by $\tilde{S}_0 = S_0$, $\tilde{S}_T = S_T/B_T$).

Definition 1.10. Given a probability measure $\mathbb{P}$ (the "physical" or "real world" probability measure), an *equivalent risk-neutral probability measure* $\mathbb{Q}^B$ is a probability measure equivalent to $\mathbb{P}$ such that the discounted price processes of all securities follow $(\mathcal{F}_t)$-martingales under $\mathbb{Q}^B$.

The risk-neutral probability measure is a special case of the wider concept of equivalent martingale measures:

Definition 1.11. Given a probability measure $\mathbb{P}$, an *equivalent martingale measure* $\mathbb{Q}^m$ with numeraire m is a probability measure equivalent to $\mathbb{P}$ such that the price processes of all securities, normalized by the numeraire process (m_t), follow $(\mathcal{F}_t)$-martingales under $\mathbb{Q}^m$.

Thus, the risk-neutral measure $\mathbb{Q}^B$ is the martingale measure with (B_t) as numeraire. It is the most frequently used martingale measure, because in the case of deterministic interest rates, (B_t) depends only on time, but not on the state, and is therefore convenient to work with.

In the one-period, two-state, two-security model, the measure $\mathbb{Q}^B$ is uniquely determined (if it exists) by

$$S_0 = \frac{q^B S^u + (1 - q^B) S^d}{B_T},$$

where $q^B = \mathbb{Q}^B\{\omega^u\}$ and $(1 - q^B) = \mathbb{Q}^B\{\omega^d\}$. Solving this equation for q^B, we get

$$\mathbb{Q}^B\{\omega^u\} = \frac{B_T S_0 - S^d}{S^u - S^d}, \quad \mathbb{Q}^B\{\omega^d\} = \frac{S^u - B_T S_0}{S^u - S^d}. \tag{1.2}$$

A very important result in mathematical finance is the following:

Theorem 1.1 (Fundamental Theorem of Asset Pricing). *There are no arbitrage opportunities in a market model $\mathcal{M}$ iff there exists an equivalent martingale measure $\mathbb{Q}^m$.*

Proofs for this result under various assumptions are given by Harrison and Pliska (1981), Taqqu and Willinger (1987), Dalang, Morton, and Willinger (1990), Schachermayer (1992), Rogers (1994) and Kabanov and Kramkov (1994) for discrete-time frameworks, and by Delbaen (1992), Schweizer (1992), Delbaen and Schachermayer (1994a,b, 1995), Klein and Schachermayer (1996) and Kabanov and Kramkov (1998) for models in continuous time.

Proposition 1.2. *In the one-period, two-state, two-security model, the conditions for existence of a martingale measure $\mathbb{Q}^B$ equivalent to $\mathbb{P}$ are*

$$S_d < (1 + r)S_0 < S_u.$$

Proof. If only one of the inequalities were violated, either investing in bonds or investing in stocks would dominate the other investment and, thus, constitute an arbitrage opportunity. □

If we extend the model by adding an additional security (e.g., a contingent claim Ψ), the price process of the additional security, normalized by the numeraire asset m_t, must also be a martingale under $\mathbb{Q}^m$ to avoid arbitrage opportunities.

The martingale (or risk-neutral) valuation approach exploits this relation by calculating the value of a contingent claim as the normalized (discounted) expected payoff of the claim, where the expectation is taken with respect to the martingale measure $\mathbb{Q}^m$ ($\mathbb{Q}^B$). Since, under $\mathbb{Q}^B$ (the most frequently used martingale measure), the expected rate of return of all (risky) securities equals the risk-free interest rate r, this is tantamount to valuing all securities as if investors actually were risk-neutral (hence the term *risk-neutral valuation*).

Remark 1.2. Note, however, that the approach rests on the assumption of the existence of a replicating portfolio. If such a replicating portfolio exists, contingent claims can be valued *as if* investors were risk-neutral, even when they are not (as is usually the case). The principle is not applicable, in general, in so-called *incomplete market models*, where there exist contingent claims which are not attainable. In this case, prices under different equivalent martingale measures (with the same numeraire!) need not be the same, and the selection of one specific measure for pricing purposes must be justified on other grounds. Here, we will only work in complete market models, where the equivalent martingale measure for a given numeraire is unique.

1.1.5 Numerical Example

Let us illustrate these results using the following example.

Example 1.1. Assume that the current stock price, S_0, is 100, and over the next time period, the stock price may either rise to $S_T^u = 110$ or decline to $S_T^d = 90$. Without loss of generality, we assume that $T = 1$. We want to price a European call option expiring at $T = 1$ with strike price $K = 100$, provided that the simple risk-free interest rate for period $[0, T]$-deposits and loans is $r = 5\%$.

Definition 1.12. A *European call option* $C(S, K, T)$ on S with strike price K and expiration time T is a contingent claim with payoff function

$$\Gamma_T(C(\cdot)) = \max(S_T - K, 0) =: (S_T - K)_+.$$

Figure 1.1 depicts this payoff function graphically. In our example, the value of the option at the end of the period, $C_T = (S_T - K)_+$, is therefore

$$C_T(\omega) = \begin{cases} C^u = 10 & \text{if } \omega = \omega^u, \\ C^d = 0 & \text{if } \omega = \omega^d. \end{cases}$$

The replicating portfolio is given by

$$V_T(\phi(\alpha_0, \beta_0)) = \begin{cases} 110\alpha_0 + 1.05\beta_0 = 10, \\ 90\alpha_0 + 1.05\beta_0 = 0. \end{cases}$$

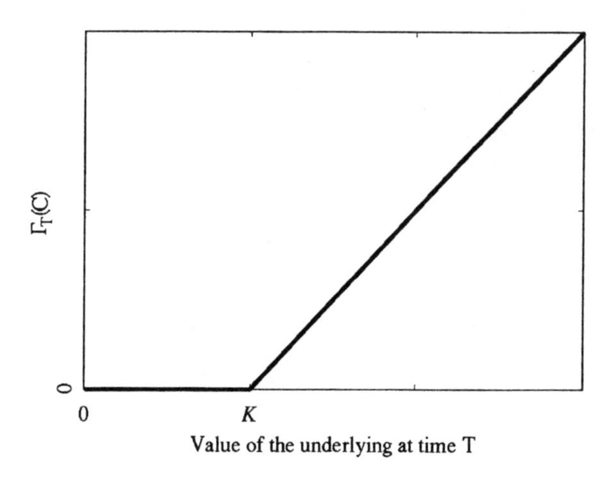

Figure 1.1: Payoff function of a European call option

The solution of this system of linear equations is $\alpha_0 = 0.5$ and $\beta_0 = -42.86$. Thus, the manufacturing cost of the European call option is given by

$$V_0(\phi) = \alpha_0 S_0 + \beta_0 = 0.5 \cdot 100 - 42.86 = 7.14.$$

For our model to be arbitrage-free, the only admissible call option price is $C_0 = V_0(\phi) = 7.14$.

Alternatively, we can arrive at this result by noting that (ruling out arbitrage) the discounted call price must follow a martingale under $\mathbb{Q}^B$. Thus, we can calculate its time $t = 0$ price as its expected value under the equivalent risk-neutral measure, discounted to the present:

$$C_0 = \frac{\mathbb{E}^B\left[(S_T - K)_+ | \mathcal{F}_0\right]}{B_T}.$$

For the numbers in our example, we get (using equation (1.2)) $q^B = 0.75$ and

$$C_0 = \left(q^B C^u + (1 - q^B) C^d\right) / B_T = \frac{0.75 \cdot 10}{1.05} = 7.14.$$

1.2 The Binomial Model

Similar to the one-period model, the binomial model (due to Cox, Ross, and Rubinstein (1979)) presented in this section will be used in Chapter 2 to introduce the idea of viewing options on stocks as compound options on the value of the firm's assets. Here, we describe the model in its basic form.

1.2.1 Model Description

In the binomial market model, the price of the risky asset (e.g., a stock) is modelled by a multiplicative binomial lattice. It is a discrete-time, finite-horizon, two-security market model. The risk-free asset (the "bond") B_t evolves according to

$$B_t = (1+r)^t \quad t = 0, 1, \ldots, T_2,$$

where T_2 denotes the model horizon. The risky security evolves as follows: Given its value at time t, its time $(t+1)$-value is either $S_{t+1} = uS_t$ or $S_{t+1} = dS_t$ with $d < 1 + r < u$ ($u, d \in \mathbb{R}_+$) and $S_0 = \underline{S}$. Thus,

$$S_{t+1}/S_t = \xi_{t+1} \in \{u, d\} \quad t = 0, 1, \ldots, T_2 - 1.$$

The ξ_t are assumed to be i.i.d. random variables on $(\Omega, \mathcal{F}, \mathbb{P})$ with

$$\mathbb{P}\{\xi_t = u\} = p = 1 - \mathbb{P}\{\xi_t = d\} \quad t = 1, 2, \ldots, T_2.$$

The stock price is then modelled as

$$S_t = S_0 \prod_{j=1}^{t} \xi_j \quad t = 1, 2, \ldots, T_2.$$

1.2.2 Replicating Portfolios

For this first exposition, we restrict ourselves to the class of path-independent contingent claims.

Definition 1.13. A contingent claim Ψ written on an underlying security S and expiring at time T is called *path-independent* if it is of the form $\Gamma(\Psi) = f(S_T)$.

One approach to pricing contingent claims in the binomial model is to construct and dynamically adjust a replicating portfolio. This can be easily seen by considering a claim expiring at some time T_1 with $1 \leq T_1 \leq T_2$. At time $(T_1 - 1)$, the stock price S_{T_1-1} together with its possible moves over the period $(T_1 - 1, T_1)$ is known, and the problem of constructing a replicating portfolio reduces to the same problem within the one-period model, which has already been discussed in Section 1.1.2. Denote the value process of the contingent claim by (Ψ_t), with $\Psi_{T_1} = \Gamma(\Psi)$. At time $(T_1 - 1)$, we are looking for a portfolio $\phi_{T_1-1} = (\alpha_{T_1-1}, \beta_{T_1-1})$ such that $V_{T_1}(\phi) = \Psi_{T_1}$. This leads to the following system of linear equations:

$$V_{T_1}(\phi) = \begin{cases} \alpha_{T_1-1} u S_{T_1-1} + \beta_{T_1-1}(1+r) = \Psi_{T_1} \\ \alpha_{T_1-1} d S_{T_1-1} + \beta_{T_1-1}(1+r) = \Psi_{T_1} \end{cases}, \tag{1.3}$$

which can be solved easily.

1.2.3 Absence of Arbitrage

At time $(T_1 - 1)$, the wealth of this portfolio, $V_{T_1-1}(\phi)$, must be the same as the value of the contingent claim, Ψ_{T_1-1}. Otherwise, arbitrage opportunities would be present. Therefore, $\Psi_{T_1-1} = V_{T_1-1}(\phi)$, and the procedure continues by constructing a replicating portfolio for the period $[T_1 - 2, T_1 - 1]$. The sequence of replicating portfolios generated in this way is also called a trading strategy.

Definition 1.14. A *trading strategy* is a sequence of portfolios

$$\phi = \phi_0, \phi_1, \ldots, \phi_{T_1-1}.$$

In other words, a trading strategy is a set of rules that tells us how to adjust our portfolio at each point in time.

Definition 1.15. A trading strategy is called *self-financing at time* t if

$$\alpha_{t-1} S_t + \beta_{t-1}(1+r) = \alpha_t S_t + \beta_t.$$

In essence, this means that at time t no funds are withdrawn or added to the portfolio.

Definition 1.16. A trading strategy is called *self-financing* (without further qualification) if the self-financing condition holds for all $t = 0, \ldots, T_1$.

By repeatedly employing the arbitrage argument given above, both the trading strategy and the option price can be determined for every $t \leq T_1$, and explicit formulae can be given (see, e.g., Musiela and Rutkowski (1997, pp. 37ff.)). Since we will use the binomial model in Chapter 2 only for illustrative purposes, we do not go into more detail here.

1.2.4 Martingale Measure

A desirable property of market models (at least from a mathematical point of view) is completeness.

Definition 1.17. A market model $\mathcal{M}$ is called *complete* if every claim Ψ is attainable in $\mathcal{M}$.

Proposition 1.3. *An arbitrage-free market model $\mathcal{M}$ is complete iff there exists (for a given numeraire m) a unique martingale measure $\mathbb{Q}^m$ for $\mathcal{M}$.*

Proof. See Musiela and Rutkowski (1997, p. 79f.). □

It can be shown that a martingale measure $\mathbb{Q}^m$ ($m \in \{B, S\}$) for the binomial model exists and is unique (for the given numeraire) iff $d < (1 + r) < u$. The single-period risk-neutral martingale probability of a price increase is given by

$$q^B = (1 + r - d)/(u - d), \tag{1.4}$$

and that of a decrease by $1 - q^B$. Using these probabilities, prices of contingent claims can be calculated by iteratively working backwards through the tree.

1.2.5 Numerical Example

Let us illustrate the pricing procedure in the binomial model using the following example.

Example 1.2. Assume that the current stock price is $S_0 = 100$, and in each period $[t-1, t]_{t \in \{1,2\}}$ it may increase to uS_{t-1} (with $u = 1.1$) or decrease to dS_{t-1} (with $d = 0.9$). The risk-free interest rate per period is $r = 0.05$. We want to value a European call option with strike price $K = 95$ and expiration time $T_1 = 2$.

We start by calculating the possible terminal stock prices S_2. For each stock price, we calculate the corresponding option value at time $T_1 = 2$ as $C_2 = (S_2 - K)_+$. For working backwards through the tree, we have two possibilities: Either construct a replicating portfolio at each node of the tree, record its value and work backwards as described above in Section 1.2.2. Alternatively, calculate the option's value at each node of the tree as its discounted expectation under the risk-neutral probability measure, record it and work backwards.

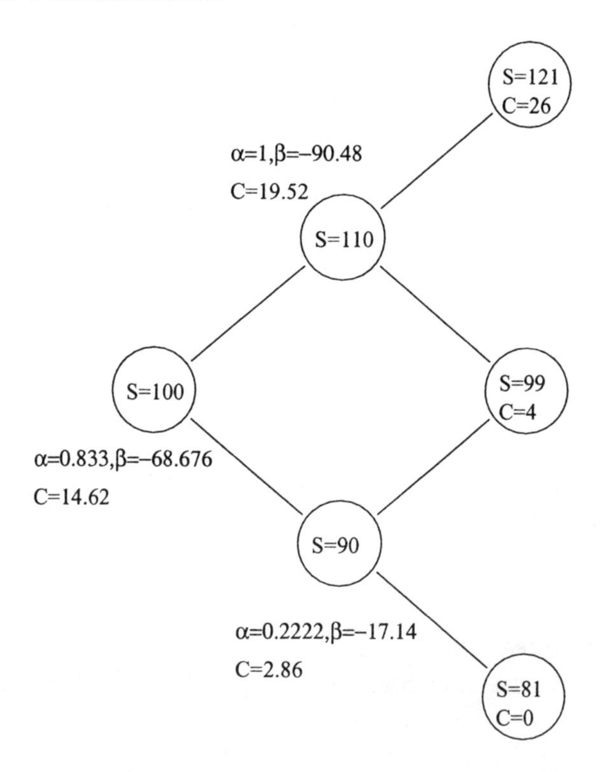

Figure 1.2: Pricing in the binomial model via replicating portfolios

We will show both alternatives here. Pricing via replicating portfolios is illustrated in Figure 1.2. Given the stock price at each node of the tree, we start with the node at which $S_1 = 110$ and try to create a replicating portfolio $\phi(\alpha_1^u, \beta_1^u)$ (where the superscript u indicates that the first change in the stock price was upwards) whose value at time $t = 2$ matches the value of the option at that time. Solving equation (1.3) yields $\alpha_1^u = 1$ and

$\beta_1^u = -90.48$. The value of the replicating portfolio (i.e., the call value) at this node is therefore $C_1^u = V_1^u(\phi) = 19.52$. For the node at which $S_1 = 90$, similar calculations yield $\alpha_1^d = 0.2222$, $\beta_1^d = -17.14$, and $C_1^d = V_1^d(\phi) = 2.86$. At time $t = 0$, we repeat this procedure and get $\alpha_0 = 0.833$, $\beta_0 = -68.676$, and $C_0 = V_0(\phi) = 14.62$.

The martingale pricing approach is shown in Figure 1.3. We use equation (1.4) to calculate the risk-neutral probabilities of a price rise or fall, respectively, and get $q^B = 0.75$ and $(1 - q^B) = 0.25$. Starting at time $t = 2$ and working backwards to the upper node at time $t = 1$, we calculate the call value at this node as the discounted expected value of the option at time $t = 2$ under $\mathbb{Q}^B$, where only nodes are considered that can be reached from the current node:

$$C_1^u = \frac{0.75 \cdot 26 + 0.25 \cdot 4}{1.05} = 19.52.$$

The corresponding calculations at the lower node at time $t = 1$ yield

$$C_1^d = \frac{0.75 \cdot 4 + 0.25 \cdot 0}{1.05} = 2.86.$$

Going back to the starting node, we get the call price from

$$C_0 = \frac{0.75 \cdot 19.52 + 0.25 \cdot 2.86}{1.05} = 14.62.$$

1.3 The Black–Scholes–Merton Model

Many firm value based option pricing models (described in Chapter 5) are built on the foundations laid down by Black and Scholes (1973) and Merton (1973). Therefore, we briefly describe the main assumptions of this famous model in its basic form (commonly referred to as the "Black–Scholes model") for reference purposes.

1.3.1 Model Description

The main difference to the models described up to now is that the Black–Scholes model is a continuous-time model. The stock price is assumed to follow a geometric Brownian motion:

$$dS_t = \mu S_t dt + \sigma S_t dW_t, \quad S_0 = \underline{S}, \tag{1.5}$$

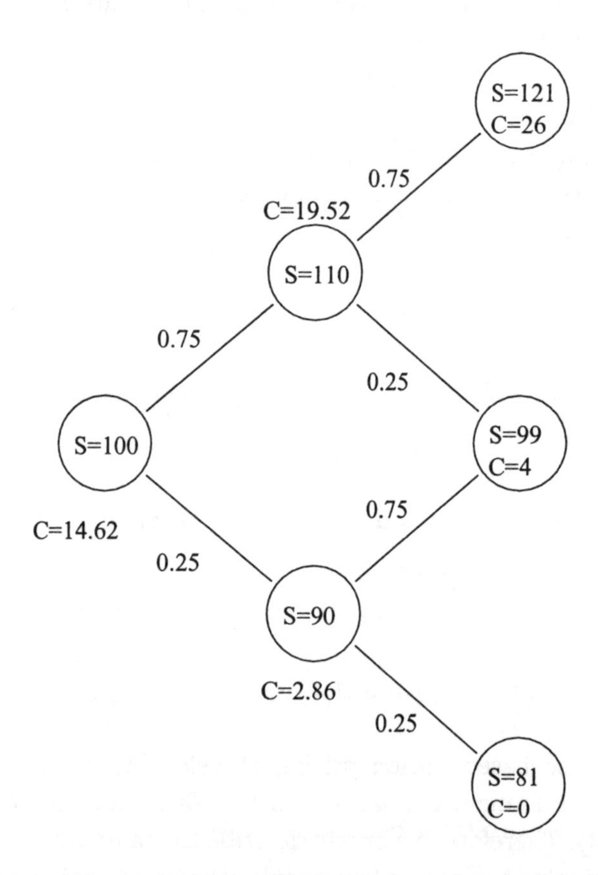

Figure 1.3: Pricing in the binomial model via risk-neutral probabilities

where W_t denotes a standard Wiener process on $(\Omega, \mathcal{F}, \mathbb{P})$.

Definition 1.18. The process $(W_t)_{\{t \geq 0\}}$ is a (simple) Brownian motion (or standard Wiener process) under $\mathbb{P}$ iff

1. (W_t) is continuous with $W_0 = \underline{W}$,
2. $(W_{s+t} - W_s) \sim N(0, \sqrt{t})$ under $\mathbb{P}$ $\quad \forall s, t \geq 0$,
3. $\mathbb{E}[(W_{s+t} - W_s)|\mathcal{F}_s] = \mathbb{E}[(W_{s+t} - W_s)|\mathcal{F}_0]$.

Thus, a Wiener process is a continuous process with independent, normally distributed increments, which do not depend on the history of the process.

In addition to the stock, there exists a riskless bond following the process

$$dB_t = rB_t dt, \quad B_0 = 1. \tag{1.6}$$

The following assumptions are made:

1. Short selling of securities is permitted.
2. There are no transactions costs or taxes.
3. All securities are perfectly divisible.
4. Trading takes place continuously.
5. The stock pays no dividends during the option's lifetime (this assumption will be relaxed in Section 1.3.5).
6. There are no arbitrage opportunities.

Some of these assumptions are only made for ease of exposition and can be relaxed. In particular, μ, r and σ need not be constant, but may be known functions of t.

1.3.2 Replicating Portfolios

Given the stock price process (S_t) and a function $\Psi(S)$, Itô's Lemma (Itô, 1951) relates a small change in the function value to a small change in the price process.

Lemma 1.1 (Itô). *Given a random variable* G *described by a stochastic differential equation of the form*

$$dG_t = A(G,t)dt + B(G,t)dW_t$$

and a function $f(G)$. *Then (suppressing subscripts and dependencies),*

$$df = \left(A\frac{\partial f}{\partial G} + \frac{B^2}{2}\frac{\partial^2 f}{\partial G^2} + \frac{\partial f}{\partial t} \right) dt + B\frac{\partial f}{\partial G}dW.$$

Proof. See Itô (1951). □

In our case here, using equation (1.5),

$$d\Psi = \left(\mu S\frac{\partial \Psi}{\partial S} + \frac{\sigma^2}{2}S^2\frac{\partial^2 \Psi}{\partial S^2} + \frac{\partial \Psi}{\partial t} \right) dt + \sigma S\frac{\partial \Psi}{\partial S}dW.$$

Suppose, we want to determine the time 0 price of an option written on S expiring at time T. The usual way in the literature to present the construction of a replicating portfolio for this option is to examine a portfolio consisting of one call option, and a fraction Δ of stocks sold short, where Δ is under the investor's discretion. Denote the value of the portfolio by Φ. Thus,

$$\begin{aligned} d\Phi &= d\Psi - \Delta dS \\ &= \left(\mu S \left(\frac{\partial \Psi}{\partial S} - \Delta \right) + \frac{\sigma^2}{2}S^2\frac{\partial^2 \Psi}{\partial S^2} + \frac{\partial \Psi}{\partial t} \right) dt \\ &\quad + \sigma S \left(\frac{\partial \Psi}{\partial S} - \Delta \right) dW_t. \end{aligned}$$

Exploiting the fact that the risk components for S and $\Psi(S)$ are perfectly correlated, we can eliminate the risk in this portfolio by choosing $\Delta = \partial\Psi/\partial S$. This *hedge ratio* Δ requires continuous adjustment (*continuous hedging*). A replicating portfolio for $\Psi(S)$ can be constructed by holding a fraction $\Delta = \partial\Psi/\partial S$ of stocks and financing the dynamic adjustment of this portfolio by borrowing money at the risk-free interest rate r.

1.3.3 Absence of Arbitrage and Risk-Neutral Pricing

As shown in the previous subsection, all risk in the portfolio Φ is eliminated by choosing (and dynamically adjusting) $\Delta = \partial\Psi/\partial S$. The absence of arbitrage requires that Φ must grow at the risk-free rate:

$$d\Phi = r\Phi dt.$$

Due to the continuous-time nature of the Black–Scholes model combined with the assumptions given above, the model is complete. From the continuous-time analogue of Proposition 1.3 and Section 1.3.2 (continuous hedging), we know that there exists a martingale measure $\mathbb{Q}^B$ such that, under this measure, the drift rate of all assets equals the risk-free interest rate. E.g., the stock price process under $\mathbb{Q}^B$ reads as

$$dS_t = rS_t dt + \sigma S_t dW_t^B, \quad S_0 = \underline{S}, \tag{1.7}$$

where dW_t^B denotes a standard Wiener process under measure $\mathbb{Q}^B$.

Derivatives can then be valued as conditional expectations under $\mathbb{Q}^B$. Given any path-independent contingent claim Ψ (i.e., $\Gamma(\Psi) = f(S_T)$), its time 0 value is calculated as

$$\Psi_0 = e^{-rT}\mathbb{E}^B[\Psi_T|\mathcal{F}_0]. \tag{1.8}$$

Black and Scholes derived their famous formulae using partial differential equations. We will show how to price derivatives in this model using the martingale approach in the sequel, since this will be our approach of choice for firm value based option pricing models in later parts of the book. In this and many later derivations in this book, we will make use of the following lemma:

Lemma 1.2. *Let (n_t) be the price process of a non-dividend-paying, continuously traded security, and let the usual assumptions of no arbitrage hold. Denote by $\mathbb{E}^B$ and $\mathbb{E}^n$ expectations with respect to the martingale measures $\mathbb{Q}^B$ and $\mathbb{Q}^n$, respectively (with (B_t) and (n_t) as numeraire processes). Then,*

$$\mathbb{E}^B[n_t \cdot I_{\{A_t\}}|\mathcal{F}_0] = \mathbb{E}^B[n_t|\mathcal{F}_0]\mathbb{E}^n[I_{\{A_t\}}|\mathcal{F}_0],$$

where A_t denotes an event which is measurable w.r.t. $\mathcal{F}_t$.

Proof. Follows directly from Corollary 2 in Geman, Karoui, and Rochet (1995, p. 449). □

Proposition 1.4 (Black–Scholes). *The value of a European call option $C(\cdot)$ on S with strike K maturing at time T is given by*

$$C_0 = S_0 N(d_1) - Ke^{-rT}N(d_2) \tag{1.9}$$

with

$$d_1 = \frac{\ln(S_0/K) + (r + \sigma^2/2)T}{\sigma\sqrt{T}} \tag{1.10}$$

and

$$d_2 = d_1 - \sigma\sqrt{T}. \tag{1.11}$$

Proof. The time T call payoff is given by $(S_T - K)_+$. Its value at time $t = 0$ (C_0) can be calculated using equation (1.8): We get

$$C_0(S, K, T) = e^{-rT}\mathbb{E}^B[(S_T - K)I_{\{S_T > K\}}] \tag{1.12}$$
$$= e^{-rT}\mathbb{E}^B[S_T I_{\{S_T > K\}}] - Ke^{-rT}\mathbb{E}^B[I_{\{S_T > K\}}] \tag{1.13}$$
$$= e^{-rT}\mathbb{E}^B[S_T]\mathbb{E}^S[I_{\{S_T > K\}}] - Ke^{-rT}\mathbb{E}^B[I_{\{S_T > K\}}] \tag{1.14}$$
$$= S_0\mathbb{Q}^S\{S_T > K\} - Ke^{-rT}\mathbb{Q}^B\{S_T > K\}, \tag{1.15}$$

where the separation of the product in the first conditional expectation, going from equation (1.13) to equation (1.14), is due to Lemma 1.2.

Thus, it only remains to derive the probabilities of the event $\{S_T > K\}$ under the martingale measures $\mathbb{Q}^B$ and $\mathbb{Q}^S$. This will be done in the next subsection, where the proof will then be completed.

1.3.4 "Measure-Independent" Derivation of Probabilities

Define a new process (X_t) on the time interval $[0, \infty)$ by

$$X_t = \frac{1}{\sigma}\ln\frac{S_t}{K}. \tag{1.16}$$

This process will have a strictly positive value at time t iff $S_t > K$. Define an associated measure $\mathbb{Q}^X$ under which (X_t) is a martingale:[3]

$$dX_t = dW_t^X.$$

From Girsanov's theorem (see, e.g., Musiela and Rutkowski (1997, p. 466)), we know that under different equivalent martingale measures $\mathbb{Q}^m$, (X_t) will be given by

$$dX_t = \mu_X^m dt + dW_t^m, \tag{1.17}$$

[3]The division by σ ensures that (X_t) is not only a martingale, but a *standard* Wiener process.

where μ_X^m denotes the drift of the process (X_t) under measure $\mathbb{Q}^m$. E.g., μ_X^B can be found easily by noting that, under $\mathbb{Q}^B$,

$$dS_t = rS_t dt + \sigma S_t dW_t^B.$$

Applying Itô's lemma to (X_t), we get

$$dX_t = \left(\frac{r - \sigma^2/2}{\sigma}\right) dt + dW_t^B,$$

so

$$\mu_X^B = \frac{r - \sigma^2/2}{\sigma}. \tag{1.18}$$

We are interested in the probability of the event $\{S_T > K\}$, which corresponds to the event $\{X_T > 0\}$. Under $\mathbb{Q}^X$,

$$X_T - X_0 = (W_T^X - W_0^X) \sim N(0, \sqrt{T}) \tag{1.19}$$

(since (W_t^X) is a standard Wiener process under $\mathbb{Q}^X$). We get further

$$\mathbb{Q}^X\{X_T > 0\} = \mathbb{Q}^X\{(X_0 + X_T - X_0) > 0\} \tag{1.20}$$

$$= \mathbb{Q}^X\{(W_T^X - W_0^X) > -X_0\}. \tag{1.21}$$

Standardizing to get an $N(0, 1)$ distributed random variable and using the symmetry of the normal distribution gives

$$\mathbb{Q}^X\{X_T > 0\} = N\left(\frac{X_0}{\sqrt{T}}\right) = N\left(\frac{\ln \frac{S_0}{K}}{\sigma\sqrt{T}}\right).$$

This gives us the probability for $\{S_T > K\}$ under $\mathbb{Q}^X$. To derive the corresponding probability for a general measure $\mathbb{Q}^m$, recall from equation (1.17) that under $\mathbb{Q}^m$,

$$dX_t = \mu_X^m dt + dW_t^m.$$

Therefore, under $\mathbb{Q}^m$,

$$X_T - X_0 = \mu_X^m T + (W_T^X - W_0^X),$$

which gives us $(X_T - X_0) \sim N(\mu_X^m, \sqrt{T})$. Following the same steps as in the case of $\mathbb{Q}^X$, we get

$$\begin{aligned} \mathbb{Q}^m\{X_T > 0\} &= \mathbb{Q}^m\{X_0 + \mu_X^m T + (W_T^m - W_0^m)\} \\ &= \mathbb{Q}^m\{(W_T^m - W_0^m) > -X_0 - \mu_X^m T\}. \end{aligned}$$

Standardizing and using the symmetry of the normal distribution again gives

$$\mathbb{Q}^m\{X_T > 0\} = N\left(\frac{X_0 + \mu_X^m}{\sqrt{T}}\right) = N\left(\frac{\ln\frac{S_0}{K}}{\sigma\sqrt{T}} + \mu_X^m\sqrt{T}\right). \tag{1.22}$$

We can find the corresponding drifts for any equivalent martingale measure $\mathbb{Q}^m$ (with traded security m as numeraire) once we know μ_X^B, by applying Theorem 1 in Geman, Karoui, and Rochet (1995, p. 448f.). We restate it here for convenience (with slight changes in notation):

Theorem 1.2 (Geman et al. (1995)). *Given a non-dividend-paying asset n_t ($n_0 = 1$) and a probability measure π equivalent to the physical measure $\mathbb{P}$ such that for any basic security without intermediate payments, the price of this security relative to n is a local martingale w.r.t. π. Let m be a non-dividend-paying numeraire such that (m_t) is a π-martingale. Then there exists a probability measure $\mathbb{Q}^m$ defined by its Radon–Nikodym derivative w.r.t. π*

$$\left.\frac{d\mathbb{Q}^m}{d\pi}\right|_{\mathcal{F}_T} = \frac{m_T}{m_0 n_T}$$

such that

1. *the basic securities are $\mathbb{Q}^m$-local martingales,*
2. *if a contingent claim has a fair price under π, then it has a fair price under $\mathbb{Q}^m$ and the hedging portfolio is the same.*[4]

Proof. See Geman, Karoui, and Rochet (1995, p. 448). □

The change of measure defined in Theorem 1.2 is a special case of Girsanov's theorem. Therefore, the Radon–Nikodym derivative in Theorem 1.2 is of the well-known form[5]

$$\left.\frac{d\mathbb{Q}^m}{d\pi}\right|_{\mathcal{F}_T} = \exp\left(\int_0^T \delta_t dW_t - \frac{1}{2}\int_0^T \delta_t^2 dt\right).$$

[4] Or, viewed alternatively, the replicating portfolio is the same.

[5] See, e.g., Musiela and Rutkowski (1997, p. 466).

δ_t is usually called the *Girsanov kernel "used to go from measure* π *to measure* $\mathbb{Q}^m$*"*. If (W_t) is a Brownian motion under π, (W_t^m) as defined by

$$W_t^m = W_t - \int_0^t \delta_s ds$$

will be a Brownian motion under Q^m. Thus, the Girsanov kernel gives us the drift of a process which is a π-Brownian motion under different martingale measures: (W_t) will have a drift of δ_t under $\mathbb{Q}^m$.

To illustrate, here we need the drift of (X_t) under $\mathbb{Q}^S$. In our setting, (X_t) is a Brownian motion under $\mathbb{Q}^X$, so $\mathbb{Q}^X$ corresponds to π in Theorem 1.2. To derive this drift, we could explicitly look for the Radon–Nikodym derivative $d\mathbb{Q}^S/d\mathbb{Q}^X$. However, since μ_X^B is already known, it will be easier to find the Girsanov kernel used to go from $\mathbb{Q}^B$ to $\mathbb{Q}^S$, and then just sum up:

$$\begin{aligned}
\frac{d\mathbb{Q}^S}{d\mathbb{Q}^B} &= \frac{S_T^B}{S_0 e^{rT}} \\
&= \frac{S_0 \exp\left((r - \sigma^2/2)T + \sigma W_T^B\right)}{S_0 e^{rT}} \\
&= \exp\left(-\frac{\sigma^2}{2}T + \sigma W_T^B\right).
\end{aligned}$$

Thus,

$$\mu_X^S = \mu_X^B + \sigma. \tag{1.23}$$

Note that the Girsanov kernels in our setting do not depend on t. Using equations (1.18) and (1.23) in equation (1.22), we have found the probabilities needed in equation (1.15) and completed the proof. □

1.3.5 Extension: Continuous Dividend Yield

If we assume that the underlying asset pays continuous (proportional) dividends at a constant rate β, equation (1.5) changes to

$$dS_t = (\mu - \beta)S_t dt + \sigma S_t dW_t, \quad S_0 = \underline{S}. \tag{1.24}$$

In this case, several changes occur in the derivation of the option pricing formula:

1. The measure change in equation (1.14) only works for non-dividend-paying assets (cf. the assumptions in Theorem 1.2). In our case, such an asset can easily be constructed by assuming that any dividends are immediately reinvested into the underlying (which is possible, given the assumptions of the model!). This artificially constructed underlying will then be a suitable numeraire.

2. μ_X^B, the drift of (X_t) under measure $\mathbb{Q}^B$, changes because of the dividend payouts to

$$\mu_X^B = \frac{r - \beta - \sigma^2/2}{\sigma}, \tag{1.25}$$

and this change carries over to μ_X^S given by equation (1.23).

3. The expectation of S_T under $\mathbb{Q}^B$ in equation (1.14) is then given by

$$\mathbb{E}^B[S_T] = e^{-\beta T} S_0 \quad \text{(instead of just } S_0\text{)}.$$

Chapter 2

Option Pricing with an Endogenous Stock Price Process

2.1 The Extended One-Period Option Pricing Model with Endogenous Stock Price Process

2.1.1 Model Description

In this section, we extend the market model described in Section 1.1. Instead of directly modelling the stock price process, we model the evolution of the firm's assets, (V_t).[1] From the asset value V_t, we *derive* the value of the company's stock endogenously. To make the model interesting from an economic point of view and to avoid trivialities, we also add another corporate security, a corporate bond. For simplicity of exposition, we assume that only one stock and one bond are outstanding.

For the moment, we assume that at the end of the period (at time T), the firm (as a whole) could be sold at a value of V_T. At the beginning of the period (at time 0), the firm issues a zerobond with a face value of D maturing at time T. Both the value process of the bond, $(D_t)_{t \in \{0,T\}}$, and that of the stock, $(S_t)_{t \in \{0,T\}}$, depend on (V_t).

(V_t) is modelled as a strictly positive $(\mathcal{F}_t)$-adapted stochastic process. $V_0 = \underline{V}$, and

$$V_T(\omega) = \begin{cases} V^u & \text{if } \omega = \omega^u \\ V^d & \text{if } \omega = \omega^d \end{cases},$$

where $V^u > V^d$. The division of the assets' terminal value between stockholder and bondholder is determined by the *absolute priority rule*, which

[1] This will be made more precise in Chapter 4.

simply states that in the case of default, equityholders only receive payments if anything is left after creditors have been paid off in full. Here, default can only occur at time T. Thus, we have the following division of assets between creditors and equityholders: If $V_T \leq D$, $D_T = V_T$ and $S_T = 0$ (default). If $V_T > D$, $D_T = D$ and $S_T = V_T - D$ (no default). The value of the stock at expiry can be written as

$$S_T = (V_T - D)_+,$$

which is, in essence, the payoff of a European call option on V_T with strike price D. As already noted by Black and Scholes (1973, p. 649f., within their continuous-time setting), in this case the bondholder effectively owns the firm's assets and has given a call option to the stockholder who can buy the assets back at time T for a strike price of D.

In addition to stocks and corporate bonds, there exists a riskless bond described by the value process $(B_t)_{t\in\{0,T\}}$ as defined in Section 1.1.1.

2.1.2 Risk-Neutral Valuation of Corporate Securities

Since in our model all corporate securities are traded, the firm's assets clearly can also be treated as a traded security (i.e., to determine the value of V_t, we simply add up the values of the firm's securities). To rule out arbitrage opportunities, the discounted values of all securities as well as their sum, $(\tilde{V}_t)$, must follow $(\mathcal{F}_t)$-martingales under $\mathbb{Q}^B$.[2]

Using equation (1.2) (and replacing S by V), we can calculate $\mathbb{Q}^B$ explicitly:

$$q^B = \mathbb{Q}^B\{\omega_u\} = \frac{B_T V_0 - V^d}{V^u - V^d}. \tag{2.1}$$

$S_0 = \tilde{S}_0$ can then be derived from V_T:

$$S_0 = \frac{\mathbb{E}^B[(V_T - D)_+|\mathcal{F}_0]}{B_T} = \frac{q^B(V^u - D)_+ + (1-q^B)(V^d - D)_+}{B_T}. \tag{2.2}$$

Similarly, the time 0 value of the corporate bond can be calculated by viewing the bond as a contingent claim with payoff $D I_{\{V_T > D\}} + V_T I_{\{V_T \leq D\}}$.

[2]This follows also from the linearity of conditional expectation.

We assume that $V^u > D$ (i.e., at least in one state, there will be no default).

$$D_0 = \frac{\mathbb{E}^B[(D \cdot I_{\{V_T>D\}} + V_T \cdot I_{\{V_T \leq D\}})|\mathcal{F}_0]}{B_T} = \tag{2.3}$$

$$= \frac{q^B D + (1-q^B)(D \cdot I_{\{V^d>D\}} + V^d \cdot I_{\{V^d \leq D\}})}{B_T}. \tag{2.4}$$

Changes in model parameters such as the risk-free interest rate or the face value of debt would lead to changes in the process (S_t). For this reason, we say that the stock price process is determined *endogenously* in this model.

2.1.3 Risk-Neutral Valuation of Options

A contingent claim Ψ expiring at time T with payoff function $\Gamma(\Psi) = f(S_T)$ and value process $(\Psi_t)_{t\in\{0,T\}}$ can be valued as

$$\Psi_0(S,\cdot) = \frac{\mathbb{E}^B[\Psi_T|\mathcal{F}_0]}{B_T} = \frac{q^B\Gamma[(V^u - D)_+] + (1-q^B)\Gamma[(V^d - D)_+]}{B_T}. \tag{2.5}$$

We see immediately from equation (2.5) that due to the endogenous nature of the stock price process, the value of the option – although formally defined in terms of the stock price – ultimately depends on the value of the firm's assets and the amount of debt outstanding.

2.1.4 Numerical Examples

Example 2.1. Assume that the current asset value is $V_0 = 100$, and possible terminal asset values are $V^u = 115.5$ and $V^d = 94.5$. The risk-free interest rate for the period is $r = 0.05$. The company has issued a discount bond with a face value of 20. We want to calculate the arbitrage-free price processes of the stock and the corporate bond.

The risk-neutral probabilities are calculated from equation (2.1):

$$q^B = \frac{1.05 \cdot 100 - 94.5}{115.5 - 94.5} = 0.5.$$

S_0 is then calculated from equation (2.2):

$$S_0 = \frac{0.5 \cdot 95.5 + 0.5 \cdot 74.5}{1.05} = 80.95,$$

and D_0 from equation (2.4):

$$D_0 = \frac{0.5 \cdot 20 + 0.5 \cdot 20}{1.05} = 19.05.$$

In this example, the corporate bond is riskless. Therefore, it yields the risk-free interest rate. To examine what happens in the case of default risk, we modify our example to the case of a highly levered firm.

Example 2.2. Same data as in Example 2.1, but $D = 100$.

Similar calculations as in the previous example yield

$$S_0 = \frac{0.5 \cdot 15.5}{1.05} = 7.38$$

and

$$D_0 = \frac{0.5 \cdot 100 + 0.5 \cdot 94.5}{1.05} = 92.62.$$

In this case, the corporate bond price is lower than it would be without default risk (100/1.05=95.24).

Example 2.3. Same data as in Example 2.1. We want to price a standard European call option on S with strike 75. Possible terminal option values are, respectively, 20.5 and 0. Using the risk-neutral valuation approach, the price of the option can be calculated as

$$C_0 = \frac{0.5 \cdot 20.5}{1.05} = 9.76.$$

2.2 The Extended Binomial Option Pricing Model with Endogenous Stock Price Process

2.2.1 Model Description

The model described in this subsection can be viewed as an extension of the standard binomial model described in Section 1.2 to allow for an endogenous stock price process. Alternatively, it can be viewed as an extension of the model described in Section 2.1 to the multi-period case.

The process for the risk-free asset, (B_t), is given by

$$B_t = (1 + r)^t \quad t = 0, 1, \ldots, T_2.$$

The asset value V_t evolves according to

$$V_{t+1}/V_t = \xi_{t+1} \in \{u, d\} \quad t = 0, 1, \ldots, T_2 - 1,$$

where $V_0 = \underline{V}$, the ξ_{t+1} are defined as in Section 1.2.1, and we still assume $d < 1 + r < u$, $(u, d \in \mathbb{R}_+)$.

At time $t = 0$, the firm issues a zerobond with a face value of D expiring at time T_2. Both the value process of this bond, $(D_t)_{t\in\{0,1,\ldots,T_2\}}$ and that of the stock, $(S_t)_{t\in\{0,1,\ldots,T_2\}}$, depend on the value process of the firm's assets, $(V_t)_{t\in\{0,1,\ldots,T_2\}}$. The division of the assets' terminal value is determined by the absolute priority rule as described in Section 2.1.1.

2.2.2 Risk-Neutral Valuation of Corporate Securities

Using the same argument as in Section 2.1.2 (all securities are traded), the discounted values of all securities as well as the discounted asset value process $(\tilde{V}_t)$ must follow $(\mathcal{F}_t)$-martingales under $\mathbb{Q}^B$.

Similar to the standard binomial model described in Section 1.2, a martingale measure for the discounted asset value process $(\tilde{V}_t)$ exists iff $d < (1 + r) < u$. The one-period martingale probability of an increase in asset value is given by (cf. equation (1.4))

$$q^B = (1 + r - d)/(u - d), \tag{2.6}$$

and that of a decrease by $(1 - q^B)$.

The processes of the asset value, stock price and corporate bond price can then be derived by iteratively working backwards through the tree using

$$S_{T_2} = (V_{T_2} - D)_+, \tag{2.7}$$

$$D_{T_2} = \min(D, V_{T_2}), \tag{2.8}$$

$$U_{t-1} = \frac{q^B U_t^u + (1 - q^B) U_t^d}{1 + r} \tag{2.9}$$

for

$$U \in \{V, S, D\}, \tag{2.10}$$

where superscripts u and d denote the up-state and down-state from the node under consideration.

In particular, we get

$$S_0 = \frac{\mathbb{E}^B\left[(V_{T_2} - D)_+ | \mathcal{F}_0\right]}{B_{T_2}}, \tag{2.11}$$

and

$$D_0 = \frac{\mathbb{E}^B\left[(D I_{\{V_{T_2} > D\}} + V_{T_2} I_{\{V_{T_2} \leq D\}}) | \mathcal{F}_0\right]}{B_{T_2}}. \tag{2.12}$$

2.2.3 Risk-Neutral Valuation of Options

Given the stock prices at all nodes of the tree, option pricing works similar to pricing in the standard binomial model. For any path-independent claim Ψ expiring at time T_1 (cf. Definition 1.13) with value process (Ψ_t) (where $\Psi_{T_1} = \Gamma_{T_1}(\Psi)$, the payoff function of the claim), we get

$$\Psi_0 = \frac{\mathbb{E}^B\left[\Psi_{T_1} | \mathcal{F}_0\right]}{B_{T_1}}. \tag{2.13}$$

Explicit formulae for equation (2.13) could be given for various standard option types. Since we use the binomial model only for introductory purposes, we will not go into more detail here.

2.2.4 Numerical Examples

Example 2.4. Assume that the current asset value is $V_0 = 100$, and in each period $[t-1, t]_{t \in \{1,2,3\}}$ it may increase to uV_{t-1} (with $u = 1.1$) or decrease to dV_{t-1} (with $d = 0.9$). The risk-free interest rate per period is $r = 0.05$. The firm has issued a discount bond maturing at $t = 3$ with a face value of 30. We want to determine the price processes of the stock and the corporate bond.

We start by calculating the possible terminal asset values V_3 and get[3]

$$V_3 = \begin{cases} 133.1 & \text{if } \xi_1 = \xi_2 = \xi_3 = u \\ 108.9 & \text{if } \exists! i \in \{1,2,3\} : \xi_i = d \\ 89.1 & \text{if } \exists! i \in \{1,2,3\} : \xi_i = u \\ 72.9 & \text{if } \xi_1 = \xi_2 = \xi_3 = d. \end{cases} \tag{2.14}$$

[3] $\exists! i$ means "there exists *exactly one* i".

From V_3, we can calculate S_3 via equation (2.7). We get

$$S_3 = \begin{cases} 103.1 & \text{if } \xi_1 = \xi_2 = \xi_3 = u \\ 78.9 & \text{if } \exists! i \in \{1,2,3\} : \xi_i = d \\ 59.1 & \text{if } \exists! i \in \{1,2,3\} : \xi_i = u \\ 42.9 & \text{if } \xi_1 = \xi_2 = \xi_3 = d. \end{cases}$$

The risk-neutral probabilities are given by equation (2.6):

$$q^B = 0.15/0.2 = 0.75.$$

Using risk-neutral valuation, we get (see equation (2.9))

$$S_2 = \begin{cases} 92.43 & \text{if } \xi_1 = \xi_2 = u \\ 70.43 & \text{if } \xi_1 \neq \xi_2 \\ 52.43 & \text{if } \xi_1 = \xi_2 = d \end{cases}, \tag{2.15}$$

$$S_1 = \begin{cases} 82.79 & \text{if } \xi_1 = u \\ 62.79 & \text{if } \xi_1 = d \end{cases}, \tag{2.16}$$

and $S_0 = 74.09$.

Since the corporate bond turns out to be riskless in this example, its value process is easily determined by $D_0 = 30/1.05^3 = 25.92$, $D_1 = 30/1.05^2 = 27.21$ and $D_2 = 30/1.05 = 28.57$.

Example 2.5. Same data as in Example 2.4. We want to value a European call option on S expiring at $t = 2$ with strike price $K = 65$.

The option prices at expiry (i.e., at time $t = 2$) are calculated using equation (2.15) to

$$C_2 = \begin{cases} 27.43 & \text{if } \xi_1 = \xi_2 = u \\ 5.43 & \text{if } \xi_1 \neq \xi_2 \\ 0 & \text{if } \xi_1 = \xi_2 = d \end{cases}. \tag{2.17}$$

The possible option values at time $t = 1$ are calculated from the results in equation (2.17) using equation (2.9):

$$C_1 = \begin{cases} 20.89 & \text{if } \xi_1 = u \\ 3.88 & \text{if } \xi_1 = d \end{cases}. \tag{2.18}$$

The option price at time $t = 0$ is then calculated as $C_0 = 15.85$.

Example 2.6. Same data as in examples 2.4 and 2.5. We want to examine the effects of a change in the capital structure of the firm on the price process of the option analyzed in Example 2.5. To this end, we assume that at time $t = 1$, the firm issues another zerobond with face value 10, maturing at time $t = 3$. To leave the asset value process (V_t) unchanged, we assume that the newly issued debt is used to repurchase shares. For simplicity, we assume that the terms of the option are not adjusted for this change in the capital structure.

Denote the price process of the original zerobond by (D_t^1) and that of the newly issued zerobond by (D_t^2). From the possible terminal asset values in equation (2.14), it is easy to see that all of the firm's debt is riskless (since $V_3 > D_3^1 + D_3^2$). Therefore, D_1^2 can be calculated simply as $10/1.05^2 = 9.07$ ($D_2^2 = 9.52$). The price process of the first zerobond, (D_t^1), remains unaffected.

The value of the stock changes at time $t = 1$, since the proceeds from the new zerobond issue are, by assumption, used to repurchase shares. The new stock prices for $t = 1, 2, 3$ can be calculated following the same steps as in Example 2.4. We get

$$S_3 = \begin{cases} 93.1 & \text{if } \xi_1 = \xi_2 = \xi_3 = u \\ 68.9 & \text{if } \exists! i \in \{1,2,3\} : \xi_i = d \\ 49.1 & \text{if } \exists! i \in \{1,2,3\} : \xi_i = u \\ 32.9 & \text{if } \xi_1 = \xi_2 = \xi_3 = d \end{cases},$$

$$S_2 = \begin{cases} 82.90 & \text{if } \xi_1 = \xi_2 = u \\ 60.90 & \text{if } \xi_1 \neq \xi_2 \\ 42.90 & \text{if } \xi_1 = \xi_2 = d \end{cases},$$

and

$$S_1 = \begin{cases} 73.72 & \text{if } \xi_1 = u \\ 53.72 & \text{if } \xi_1 = d \end{cases}.$$

The corresponding new option prices are given by

$$C_2 = \begin{cases} 17.90 & \text{if } \xi_1 = \xi_2 = u \\ 0 & \text{if } \xi_1 \neq \xi_2 \\ 0 & \text{if } \xi_1 = \xi_2 = d \end{cases},$$

and

$$C_1 = \begin{cases} 12.79 & \text{if } \xi_1 = u \\ 0 & \text{if } \xi_1 = d \end{cases}.$$

Thus, if $\xi_1 = u$, the option price would change from 20.89 (cf. equation (2.18)) to 12.79, and if $\xi_1 = d$, the option price would change from 3.88 to 0. These option price changes are solely due to the change in the firm's capital structure.[4] It is this direct link between (changes in) the firm's capital structure and derivatives pricing that Toft and Prucyk (1997, p. 1151) identified to have been "with few exceptions, [...] ignored in the theoretical and empirical option pricing literature", and that will be the central topic in this book.

2.3 The Extended Black–Scholes Model with Endogenous Stock Price Process

2.3.1 Model Description

Here, we assume that the value of the firm's assets follows a geometric Brownian motion:

$$dV_t = \mu_V V_t dt + \sigma_V V_t dW_t, \quad V_0 = \underline{V},$$

where the subscript V in μ_V and σ_V serves to clarify that these are the parameters of the asset value process (V_t). There is only one class of equity securities outstanding, and for simplicity of exposition, we assume that equity consists of only one share of common stock. The firm has issued a zerobond with face value D expiring at time T_2. The division of the assets' terminal value is determined by the absolute priority rule (see page 23). Furthermore, there exists a riskless bond as defined in equation (1.6), and the Black–Scholes assumptions given on page 15 hold.

2.3.2 Risk-Neutral Valuation of Corporate Securities

Like the standard Black–Scholes model, the extended Black–Scholes model is complete because of the possibility of continuous trading (see Section

[4]Exchange-traded stock options are not protected against future changes in a firm's capital structure.

1.3.2). By Proposition 1.3, there exists a (unique) equivalent martingale measure $\mathbb{Q}^B$ with numeraire process (B_t).

The problem of pricing corporate securities in this setting was first studied in detail by Merton (1974), although the principles have already been laid down by Black and Scholes (1973). Using the approach common at this time, Merton arrived at his results via partial differential equations, whereas we will use the martingale approach here. Under the equivalent risk-neutral measure, price processes of all securities have a drift rate equal to the risk-free interest rate. Since the asset value V is just the sum of the stock and the bond, it also grows at the risk-free rate r under $\mathbb{Q}^B$:

$$dV_t = rV_t dt + \sigma_V V_t dW_t^B.$$

2.3.2.1 Equity

Absolute priority implies that at maturity of the zerobond (at time T_2), the stockholder decides whether to pay off the bondholder and receive the assets, or to default and give the assets to the bondholder. His payoff at time T_2 therefore resembles that of a standard European call option on V:

$$S_{T_2} = (V_{T_2} - D)_+ \tag{2.19}$$

Since the asset value follows a geometric Brownian motion, we get the value of equity simply by replacing S_0, K and T in equation (1.9) by V, D and T_2, respectively. The stock return volatility σ in equations (1.5) and (1.10) is replaced by the volatility of the asset value, σ_V. Thus, we arrive at

$$S_0 = V_0 N(d_1) - De^{-rT_2} N(d_2) \tag{2.20}$$

with

$$d_1 = \frac{\ln(V_0/D) + (r + \sigma_V^2/2)T_2}{\sigma_V \sqrt{T_2}}$$

and

$$d_2 = d_1 - \sigma_V \sqrt{T_2}.$$

2.3.2.2 Corporate Bond

As noted by Merton (1974, p. 454), the value of the corporate bond is just the difference between V_0 and S_0:

$$D_0 = V_0 - S_0 = V_0 N(-d_1) + De^{-rT_2} N(d_2). \tag{2.21}$$

Adding up again, it is easy to see that the first terms on the right hand sides of equations (2.20) and (2.21) sum to V_0, and the remaining terms cancel out.

2.3.3 Risk-Neutral Valuation of Options

The problem of valuing options on options (or "compound call options") was first studied by Geske (1977, 1979). Here, we re-derive his compound call option valuation formula using the martingale pricing method.

Proposition 2.1 (Geske (1979)). *In the model described in this section, the value of a European call option on* S *with strike* K *expiring at time* T_1 *is given by*

$$\begin{aligned} C(S,K,T_1) = {} & V_0 N(h + \sigma\sqrt{T_1}, k + \sigma\sqrt{T_2}, \sqrt{T_1/T_2}) \\ & - e^{-rT_2} D N(h, k, \sqrt{T_1/T_2}) - e^{-rT_1} K N(h) \end{aligned} \tag{2.22}$$

with

$$h = \frac{\ln(V_0/\overline{V}) + (r - \sigma^2/2)T_1}{\sigma\sqrt{T_1}}, \tag{2.23}$$

$$k = \frac{\ln(V_0/D) + (r - \sigma^2/2)T_2}{\sigma\sqrt{T_2}}, \tag{2.24}$$

$$\overline{V}: \ C\left(\overline{V}, D, (T_2 - T_1)\right) = K. \tag{2.25}$$

Proof. At time T_1, the call will be exercised iff it is in the money, i.e., iff $S_{T_1} > K$. Since, in this model, the stock itself can be viewed as a call on V with strike D maturing at time T_2 (see previous subsection), the event $\{S_{T_1} > K\}$ can be re-expressed in terms of V, the state variable of the model:

$$\{S_{T_1} > K\} \Leftrightarrow \{V_{T_1} > \overline{V}\},$$

where $\overline{V}$ is defined in equation (2.25) and denotes the critical value of V_{T_1} (i.e., for values of V_{T_1} above $\overline{V}$, the call would be exercised).

As usual, the time 0 value of the option can be calculated as its discounted expected value under $\mathbb{Q}^B$:

$$\begin{aligned} C(S,K,T_1) &= e^{-rT_1}\mathbb{E}^B[(S_{T_1}-K)I_{\{S_{T_1}>K\}}] \\ &= e^{-rT_1}\mathbb{E}^B[S_{T_1}I_{\{S_{T_1}>K\}}] - e^{-rT_1}K\mathbb{E}^B[I_{\{S_{T_1}>K\}}] \\ &= e^{-rT_1}\mathbb{E}^B[C_{T_1}(V_{T_1},D,T_2)I_{\{V_{T_1}>\overline{V}\}}] - e^{-rT_1}K\mathbb{Q}^B\{V_{T_1}>\overline{V}\} \\ &= e^{-rT_1}\Big(\Big(\mathbb{E}^B[V_{T_1}]\mathbb{E}^V[I_{\{V_{T_2}>D\}}] \\ &\quad -e^{-r(T_2-T_1)}D\mathbb{E}^B[I_{\{V_{T_2}>D\}}]\Big)I_{\{V_{T_1}>\overline{V}\}} - K\mathbb{Q}^B\{V_{T_1}>\overline{V}\}\Big) \\ &= V_0\mathbb{Q}^V\{V_{T_2}>D, V_{T_1}>\overline{V}\} \\ &\quad - e^{-rT_2}D\mathbb{Q}^B\{V_{T_2}>D, V_{T_1}>\overline{V}\} - e^{-rT_1}K\mathbb{Q}^B\{V_{T_1}>\overline{V}\}. \end{aligned} \tag{2.26}$$

Thus, it only remains to derive the probabilities for the event $\{V_{T_2} > D, V_{T_1} > \overline{V}\}$ under measures $\mathbb{Q}^B$ and $\mathbb{Q}^V$. This will be done in the next subsection, where the proof will then be completed.

2.3.4 "Measure-independent" Derivation of Probabilities

We will derive the probabilities for the event $\{V_{T_2} > D, V_{T_1} > \overline{V}\}$ independently of any specific martingale measure (or, equivalently, for a general martingale measure $\mathbb{Q}^m$), as a function of the Girsanov kernel associated with the respective measures.

Define a new process (X_t) on the time interval $[0,\infty)$ by

$$X_t = \frac{1}{\sigma}\ln\frac{V_t}{V_0}.$$

Define further the "normalized critical firm value"

$${}_X\overline{V} = \frac{1}{\sigma}\ln\frac{\overline{V}}{V_0}$$

and the "normalized debt principal"

$${}_XD = \frac{1}{\sigma}\ln\frac{D}{V_0}.$$

The event $\{V_{T_2} > D, V_{T_1} > \overline{V}\}$ is then equivalent to the event $\{X_{T_2} > {}_XD, X_{T_1} > {}_X\overline{V}\}$. Under $\mathbb{Q}^X$, (X_t) is a Wiener process. Therefore,

$$X_T - X_0 = (W_T^X - W_0^X) \sim N(0,\sqrt{T}) \quad \text{under } \mathbb{Q}^X.$$

Standardizing and using the symmetry of the normal distribution, we get

$$
\begin{aligned}
\mathbb{Q}^X\{X_{T_2} > x_D, X_{T_1} > x_{\overline{V}}\} &= N\left(-\frac{\ln\frac{D}{V_0}}{\sigma\sqrt{T_2}}, -\frac{\ln\frac{\overline{V}}{V_0}}{\sigma\sqrt{T_1}}, \sqrt{\frac{T_1}{T_2}}\right) \\
&= N\left(\frac{\ln\frac{V_0}{D}}{\sigma\sqrt{T_2}}, \frac{\ln\frac{V_0}{\overline{V}}}{\sigma\sqrt{T_1}}, \sqrt{\frac{T_1}{T_2}}\right).
\end{aligned}
$$

Under measures $\mathbb{Q}^m$ with $m \in \{B, V\}$,

$$
dX_t = \mu_X^m + dW_t^m,
$$

which implies that, under $\mathbb{Q}^m$,

$$
X_T - X_0 = (\mu_X^m T + (W_T^m - W_0^m)) \sim N(\mu_X^m T, \sqrt{T}).
$$

We get further

$$
\begin{aligned}
\mathbb{Q}^m\{\cdot\} &= N\left(-\left(\frac{\ln\frac{D}{V_0}}{\sigma\sqrt{T_2}} - \mu_X^m\sqrt{T_2}\right), -\left(\frac{\ln\frac{\overline{V}}{V_0}}{\sigma\sqrt{T_1}} - \mu_X^m\sqrt{T_1}\right), \sqrt{\frac{T_1}{T_2}}\right) \\
&= N\left(\left(\frac{\ln\frac{V_0}{D}}{\sigma\sqrt{T_2}} + \mu_X^m\sqrt{T_2}\right), \left(\frac{\ln\frac{V_0}{\overline{V}}}{\sigma\sqrt{T_1}} + \mu_X^m\sqrt{T_1}\right), \sqrt{\frac{T_1}{T_2}}\right).
\end{aligned}
\tag{2.27}
$$

The drifts μ_X^m needed in equation (2.27) are given in equations (1.18) and (1.23) (note that the drifts are the same as in Section 1.3.4, although (X_t) was defined slightly differently there – the drifts do not depend on the constant used for standardizing). Together with equation (2.26), the proof is completed. □

2.3.5 Numerical Examples

Example 2.7. Assume that, under the physical measure $\mathbb{P}$, the asset value follows a geometric Brownian motion with $\sigma_V = 0.5$. The current asset value is $V_0 = 100$, and the risk-free interest rate is $r = 0.05$. For simplicity of exposition, we assume that only one share of common stock is outstanding. The firm has issued a zerobond maturing at time $T_2 = 3$ with a face value of 30. We want to determine the time 0 prices of the stock and the corporate bond.

Regardless of μ_V, under the martingale measure $\mathbb{Q}^B$, the asset value (as the sum of traded securities) follows a geometric Brownian motion with drift rate r:

$$dV_t = 0.05V_t + 0.5V_t dW_t^B.$$

The stock price S_0 is given by equation (2.20):

$$S_0 = 100N(d_1) - 30e^{-0.05 \cdot 3}N(d_2) \tag{2.28}$$

with

$$d_1 = \frac{\ln(100/30) + 3(0.05 + 0.25/2)}{0.5\sqrt{3}} \tag{2.29}$$

and

$$d_2 = d_1 - 0.5\sqrt{3}. \tag{2.30}$$

We get a stock price of 75.22.

The bond can either be valued using equation (2.21), or by subtracting the stock price from the asset value:

$$D_0 = 100N(-d_1) + 30e^{-0.05 \cdot 3}N(d_2) = 100 - 75.22 = 24.78.$$

Example 2.8. Same data as in Example 2.7. We want to value a European call option on S expiring at $T_1 = 2$ with strike $K = 60$.

The value of this call, although formally defined in terms of the value of S, ultimately depends on the value of V_2, i.e., the asset value at maturity of the option. It is given by equations (2.22) through (2.25) (we denote the event $\{V_{T_2} > D, V_{T_1} > \overline{V}\}$ with $\{A_{T_2}\}$):

$$\begin{aligned} C(75.22, 60, 2) &= C^{(1)}(C^{(2)}(100, 30, 3), 60, 2) \\ &= 100\mathbb{Q}^V\{A_{T_2}\} - 30e^{-0.05 \cdot 3}\mathbb{Q}^B\{A_{T_2}\} - 60e^{-0.05 \cdot 2}\mathbb{Q}^B\{V_{T_1} > \overline{V}\} \end{aligned}$$

with

$$\mathbb{Q}^V\{A_{T_2}\} = N\left(\frac{\ln\frac{100}{88.437} + 2(0.05 + \frac{0.25}{2})}{0.5\sqrt{2}}, \frac{\ln\frac{100}{30} + 3(0.05 + \frac{0.25}{2})}{0.5\sqrt{3}}, \sqrt{\frac{2}{3}}\right),$$

$$\mathbb{Q}^B\{A_{T_2}\} = N\left(\frac{\ln\frac{100}{88.437} + 2(0.05 - \frac{0.25}{2})}{0.5\sqrt{2}}, \frac{\ln\frac{100}{30} + 3(0.05 - \frac{0.25}{2})}{0.5\sqrt{3}}, \sqrt{\frac{2}{3}}\right),$$

and

$$\mathbb{Q}^B\{V_{T_1} > \overline{V}\} = N\left(\frac{\ln \frac{100}{88.437} + 2(0.05 - \frac{0.25}{2})}{0.5\sqrt{2}}\right).$$

The critical value $\overline{V}$ that separates the exercise- and non-exercise-regions of the first option (the call on the stock) is calculated using equation (2.25):

$$\overline{V}: \; C(\overline{V}, 30, 1) = 60 \quad \Rightarrow \overline{V} = 88.437.$$

We get a call option price $C_0 = 36$.

Example 2.9. Same data as in Examples 2.7 and 2.8. We want to examine the effects of a change in the capital structure of the firm on the price of the option analyzed in Example 2.8. To this end, we assume that immediately after time $t = 0$, the firm issues another zerobond (ranking equally with the first zerobond in case of default) with face value 10, maturing at time $t = 3$. To leave the asset value process (V_t) unchanged, we assume that the newly issued debt is used to repurchase a fraction of the share. For simplicity, we assume that the terms of the option are not adjusted for this change in the capital structure.

Immediately after issuing the second zerobond, the firm has outstanding zerobonds with a total face value of 40, maturing at time $t = 3$. Therefore, assuming repurchase of a fraction of the stock, its value drops from 75.22 (calculated in Example (2.7)) to 68.08, given by:

$$S_0 = 100N(d_1) - 40e^{-0.05 \cdot 3}N(d_2)$$

with

$$d_1 = \frac{\ln(100/40) + 3 \cdot (0.05 + 0.25/2)}{0.5\sqrt{3}}$$

and

$$d_2 = d_1 - 0.5\sqrt{3}.$$

The price of the option analyzed in Example 2.8 can be calculated from

$$\begin{aligned} C(68.08, 60, 2) &= C^{(1)}(C^{(2)}(100, 40, 3), 60, 2) \\ &= 100\mathbb{Q}^V\{A_{T_2}\} - 40e^{-0.05 \cdot 3}\mathbb{Q}^B\{A_{T_2}\} \\ &\quad - 60e^{-0.05 \cdot 2}\mathbb{Q}^B\{V_{T_1} > \overline{V}\} \end{aligned}$$

with

$$\mathbb{Q}^V\{A_{T_2}\} = N\left(\frac{\ln\frac{100}{97.708} + 2(0.05 + \frac{0.25}{2})}{0.5\sqrt{2}}, \frac{\ln\frac{100}{40} + 3(0.05 + \frac{0.25}{2})}{0.5\sqrt{3}}, \sqrt{\frac{2}{3}}\right),$$

$$\mathbb{Q}^B\{A_{T_2}\} = N\left(\frac{\ln\frac{100}{97.708} + 2(0.05 - \frac{0.25}{2})}{0.5\sqrt{2}}, \frac{\ln\frac{100}{40} + 3(0.05 - \frac{0.25}{2})}{0.5\sqrt{3}}, \sqrt{\frac{2}{3}}\right),$$

and

$$\mathbb{Q}^B\{V_T > \overline{V}\} = N\left(\frac{\ln\frac{100}{97.708} + 2(0.05 - \frac{0.25}{2})}{0.5\sqrt{2}}\right),$$

where the critical value $\overline{V}$ that separates the exercise- and non-exercise-regions of the first option changes from 88.437 (in Example 2.8) to:

$$\overline{V} := C(\overline{V}, 40, 1) = 60 \quad \Rightarrow \overline{V} = 97.708.$$

The change in the capital structure leads to a decrease in the option price from 36 (in Example 2.8, with $D = 30$) to 32.10 (retaining the assumption that the option strike is not adjusted for the share repurchase).

Chapter 3

Exotic Options

In this chapter, we will give a brief introduction on some exotic options. We describe only a small selection of exotics that will form the basis for many of the firm value based models described in the following chapters.

We will distinguish between two fundamentally different types of exotic options: Those whose payoffs depend on whether the underlying has reached a certain barrier before the option's maturity (*barrier options*), and those where such a dependence does not exist (*non-barrier options*). We start with the class of non-barrier options, which is mathematically easier to handle.

3.1 Non-Barrier Exotic Options

From the non-barrier class, we will only need the heaviside option in later chapters:

Definition 3.1. A *European heaviside option* $H(S, K, T)$ on S with strike price K and expiration time T is a contingent claim with payoff $\Gamma_T(H(\cdot)) = I_{\{S_T > K\}}$.

Figure 3.1 gives a graphical representation of the heaviside payoff at maturity T.

Risk-neutral valuation gives us

$$\begin{aligned} H(S, K, T) &= e^{-rT}\mathbb{E}^B[I_{\{S_T > K\}}] \\ &= e^{-rT}\mathbb{Q}^B\{S_T > K\}. \end{aligned}$$

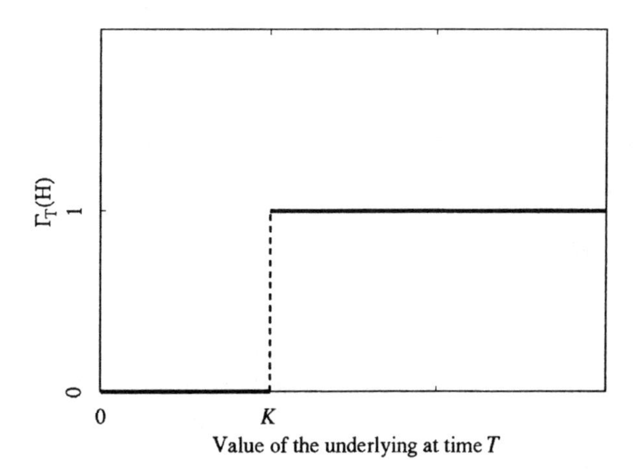

Figure 3.1: Payoff function of a European heaviside option

This probability has already been found in Section 1.3.4. Combining equations (1.22) and (1.18), we get

$$H(S, K, T) = e^{-rT} N(d_2), \tag{3.1}$$

where d_2 is defined as in equation (1.11). The heaviside is just equal to the "second half" of $1/K$ standard call options.

3.2 Barrier Options

Good introductions to barrier option pricing can be found, e.g., in Björk (1998, chapter 13) and Zhang (1998, chapters 10 and 11). A comprehensive overview of pricing formulas can be found in Rich (1994). Most of the literature on barrier options proceeds by first deriving the first passage time distribution and the restricted density of the "truncated" underlying process under the measure $\mathbb{Q}^B$. Then, derivatives are priced by integrating this truncated density function. Care has to be taken to adjust the integration domain appropriately.

We will take a different approach here and derive the corresponding probabilities "independently" of any specific martingale measure (or, equivalently, for a general equivalent martingale measure $\mathbb{Q}^m$, as opposed to only for the risk-neutral measure $\mathbb{Q}^B$). Following the derivation of these probabilities, we will show how this approach relates to the "decomposition-style" approach taken, e.g., by Björk (1998, chapter 13). We note that the probabilities derived in the sequel can be found, e.g., in Rich (1994) (although not in a form as accessible and explicit as here, especially for measures other than $\mathbb{Q}^B$). The main purpose of this section is to introduce the probabilistic approch used in later chapters within a comparatively simple setting.

3.2.1 Preliminaries

Definition 3.2. For any $y \in \mathbb{R}$, the *hitting time* or *first passage time* of a stochastic process (S_t) to y is defined by

$$\tau = \inf\{t \geq 0 | S_t = y\}.$$

Definition 3.3. Given a stochastic process (S_t), the process $(S_{\min\{t,\tau\}})$ is called the *absorbed process*.

A well-known result about Brownian motions "reflected" at barriers is the so-called *Reflection Principle*:

Lemma 3.1 (Reflection Principle). *Given a Brownian motion (W_t) (under $\mathbb{P}$), starting at $W_0 > L$, and a constant $b \in \mathbb{R}$. Denote the first passage time of (W_t) to L by τ. Then,*

$$\mathbb{P}\{W_T > b, \tau \in [0, T]\} = \mathbb{P}\{W_T < 2L - b\}.$$

Proof. Follows directly from the Markov property of Brownian motion. Intuitively, conditional on touching the barrier at $\tau < T$, paths that end up above b and paths that end up below $2L - b$ have equal probability of occurrence, since the distance from W_0 is at least b in both cases. For a more precise treatment, see e.g. Harrison (1985). □

Remark 3.1. The main benefit of the Reflection Principle is to allow the expression of the probability of an event which is conditional on a barrier as an unconditional probability. Note that equivalent statements can be made for the case of an up-barrier (when $W_0 < L$).

3.2.2 Option Types and Barrier-Dependent Probabilities

We start this section with a definition of four basic barrier option types.

Definition 3.4. A *down-and-in option with barrier* $L(\cdot)$ is an option which begins to exist as soon as the underlying asset reaches the barrier for the first time (from above).

Definition 3.5. An *up-and-in option with barrier* $L(\cdot)$ is an option which begins to exist as soon as the value of the underlying asset reaches the barrier for the first time (from below).

Definition 3.6. An *up-and-out option with barrier* $L(\cdot)$ is an option which ceases to exist as soon as the value of the underlying asset reaches the barrier for the first time (from below).

Definition 3.7. A *down-and-out option with barrier* $L(\cdot)$ is an option which ceases to exist as soon as the value of the underlying asset reaches the barrier for the first time (from above).

For each of these option types, there exist corresponding barrier-dependent probabilities. An example would be the probability of the event $\{V_T > \overline{V}, \tau \notin [0,T]\}$, i.e., the value of the underlying at time T exceeds the prespecified threshold $\overline{V}$, and the process has not touched the barrier up to and including time T.

Although the following result is well-known, we show here in detail the derivation of one of the probabilities that we will need later on to illustrate the approach taken here. In later chapters, this will be extended to more complicated events.

Lemma 3.2 (">/down-and-out"). *Define the default process* (X_t) *on the time interval* $[0, \infty)$ *as follows:*

$$X_t = \frac{1}{\sigma} \ln \frac{V_t}{l_t}, \tag{3.2}$$

where (V_t) *is a geometric Brownian motion under* $\mathbb{P}$, $l_t = l_0 e^{\rho t}$, *and* $V_0 > l_0$ *(i.e., we are considering a down-barrier here). Let* τ *denote the first passage time of the default process to 0 for* $t > 0$. *Given a prespecified threshold* $\overline{V}$, *the event* $\{V_T > \overline{V}, \tau \notin [0,T]\}$ *is equivalent to the event* $\{X_T > {}_X\overline{V}, \tau \notin [0,T]\}$ *with* ${}_X\overline{V} = \frac{1}{\sigma}\ln\frac{\overline{V}}{l_T}$ *(the normalized threshold). The* $\mathbb{Q}^m$*–probability for* $\{V_T > \overline{V}, \tau \notin [0,T]\}$ *is given by*

$$\begin{aligned} \mathbb{Q}^m\{V_T > \overline{V}, \tau \notin [0,T]\} = {} & N\left(\frac{\mu_X^m T + X_0 - {}_X\overline{V}}{\sqrt{T}}\right) \\ & - e^{-2\mu_X^m X_0} N\left(\frac{\mu_X^m T - X_0 - {}_X\overline{V}}{\sqrt{T}}\right). \end{aligned} \tag{3.3}$$

If $\overline{V} < l_T$*, set* $\overline{V} = l_T$ *in equation* (3.3).

Proof. Assume for the moment that $\rho = 0$ (i.e., the barrier is constant). As a first step,[1] we will derive the $\mathbb{Q}^X$-probability:

$$\begin{aligned} \mathbb{Q}^X\{X_T > {}_X\overline{V}, \tau \notin [0,T]\} &= \mathbb{Q}^X\{X_T > {}_X\overline{V}\} - \mathbb{Q}^X\{X_T > {}_X\overline{V}, \tau \leq T\} \\ &= \mathbb{Q}^X\{X_T < 2X_0 - {}_X\overline{V}\} - \mathbb{Q}^X\{X_T < -{}_X\overline{V}\} \\ &= N\left(\frac{X_0 - {}_X\overline{V}}{\sqrt{T}}\right) - N\left(-\frac{({}_X\overline{V} + X_0)}{\sqrt{T}}\right). \end{aligned}$$

Remember that for $\overline{V} < l_T$, $\overline{V}$ must be replaced with l_T.

Switching to integral form (densities), this reads as

$$\mathbb{Q}^X\{X_T > {}_X\overline{V}, \tau \notin [0,T]\} = \frac{1}{\sqrt{2\pi}}\left[\int_{\frac{{}_X\overline{V} - X_0}{\sqrt{T}}}^{\infty} e^{-z^2/2}dz - \int_{\frac{{}_X\overline{V} + X_0}{\sqrt{T}}}^{\infty} e^{-z^2/2}dz\right].$$

Changing integration variables from standard normal to dw_T^X (where w_t^X denotes a specific realization of the process (W_t^X)) gives

$$\begin{aligned} \mathbb{Q}^X\{X_T > {}_X\overline{V}, \tau \notin [0,T]\} = \frac{1}{\sqrt{2\pi T}} & \left[\int_{{}_X\overline{V} - X_0}^{\infty} e^{-\left(\frac{w_T^X}{\sqrt{T}}\right)^2/2} dw_T^X \right. \\ & \left. - \int_{{}_X\overline{V} - X_0}^{\infty} e^{-\left(\frac{w_T^X + 2X_0}{\sqrt{T}}\right)^2/2} dw_T^X\right]. \end{aligned}$$

Thus,

$$\mathbb{Q}^X\{X_T \in dw_T^X, \tau \notin [0,T]\} = \int_{{}_X\overline{V} - X_0}^{\infty} f(0, \sqrt{T})dw_T^X - \int_{{}_X\overline{V} - X_0}^{\infty} f(-2X_0, \sqrt{T})dw_T^X,$$

[1] This is basically the univariate version of the proof of Lemma 6.1 (Proposition 3 in Reneby (1998)).

where $f(\mu, \sigma)$ denotes the density of a normal distribution with mean μ and standard deviation σ. For $\mathbb{Q}^m\{V_T > \overline{V}, \tau \notin [0,T]\}$, we get from the definition of the Radon–Nikodym derivative (see, e.g., Baxter and Rennie (1996, pp. 63ff.)):

$$\begin{aligned}
\mathbb{Q}^m\{X_T > {}_X\overline{V}, \tau \notin [0,T]\} &= \int_{{}_X\overline{V}-X_0}^{\infty} R^{X\to m}\mathbb{Q}^X\{X_T \in dw_T^X, \tau \notin [0,T]\}dw_T^X \\
&= e^{-\frac{1}{2}(\mu_X^m)^2 T}\left[\int_{{}_X\overline{V}-X_0}^{\infty} e^{\mu_X^m w_T^X} f(0,\sqrt{T})dw_T^X \right. \\
&\quad \left. - \int_{{}_X\overline{V}-X_0}^{\infty} e^{\mu_X^m w_T^X} f(-2X_0,\sqrt{T})dw_T^X\right].
\end{aligned}$$

Square completion and cancellation of terms gives

$$\begin{aligned}
\mathbb{Q}^m\{X_T > {}_X\overline{V}, \tau \notin [0,T]\} &= \int_{{}_X\overline{V}-X_0}^{\infty} f(\mu_X^m T, \sqrt{T})dw_T^X \\
&\quad - e^{-2\mu_X^m X_0}\int_{{}_X\overline{V}-X_0}^{\infty} f(-2X_0 + \mu_X^m T, \sqrt{T})dw_T^X.
\end{aligned}$$

Changing integration variables to standard normal again, we get

$$\begin{aligned}
\mathbb{Q}^m\{X_T > {}_X\overline{V}, \tau \notin [0,T]\} &= \int_{\frac{{}_X\overline{V}-\mu_X^m T - X_0}{\sqrt{T}}}^{\infty} f(0,1)dz \\
&\quad - e^{-2\mu_X^m X_0}\int_{\frac{{}_X\overline{V}-\mu_X^m T + X_0}{\sqrt{T}}}^{\infty} f(0,1)dz \\
&= N\left(\frac{\mu_X^m T + X_0 - {}_X\overline{V}}{\sqrt{T}}\right) \\
&\quad - e^{-2\mu_X^m X_0} N\left(\frac{\mu_X^m T - X_0 - {}_X\overline{V}}{\sqrt{T}}\right).
\end{aligned}$$

For the case of an exponential barrier, note that (following Zhang (1998, p. 262)) the probability that (V_t) hits the exponential barrier $l_t = l_0 e^{\rho t}$ in the interval $[0,T]$ equals the probability that $(e^{-\rho t}V_t)$ hits the constant barrier $L = l_0$ in this time interval. Repeating the previous steps applying this insight yields the result for the exponential barrier case. □

Although working directly on the standardized variables X_0 and ${}_X\overline{V}$ is often more convenient and less error-prone, some people might prefer the probabilities expressed in terms of the original variables. In the sequel, we show how these representations can be derived from our lemmata, using the ">/down-and-out"-probability as an example. This also serves to show the link between our representation and that found in other published work.

For the case of a constant barrier, we get the following corollary:

Corollary 3.1. *The $\mathbb{Q}^m$-probability that the value of V_T exceeds some prespecified threshold $\overline{V}$ without the process hitting a constant barrier L in the time interval $[0,T]$ is given by*

$$\begin{aligned}\mathbb{Q}^m\{V_T > \overline{V}, \tau \notin [0,T]\} &= N\left(\frac{\ln\frac{V_0}{\overline{V}}}{\sigma\sqrt{T}} + \mu_X^m\sqrt{T}\right)\\ &\quad - \left(\frac{L}{V_0}\right)^{\alpha} N\left(\frac{\ln\frac{L^2}{V_0\overline{V}}}{\sigma\sqrt{T}} + \mu_X^m\sqrt{T}\right)\end{aligned} \tag{3.4}$$

with

$$\alpha = \frac{2\mu_X^m}{\sigma}. \tag{3.5}$$

If $\overline{V} < l_T$, set $\overline{V} = l_T$ when calculating this probability.

Proof. Follows directly from Lemma 3.2. □

For the case of an exponential barrier ($l_t = l_0 e^{\rho t}$), we get the following corollary:

Corollary 3.2. *The $\mathbb{Q}^m$-probability that the value of V_T exceeds some prespecified threshold $\overline{V}$ without the process hitting l_t before is given by*

$$\begin{aligned}\mathbb{Q}^m\{V_T > \overline{V}, \tau \notin [0,T]\} &= N\left(\frac{\ln\frac{V_0}{\overline{V}} + \rho T}{\sigma\sqrt{T}} + \mu_X^m\sqrt{T}\right)\\ &\quad - \left(\frac{L}{V_0}\right)^{\alpha} N\left(\frac{\ln\frac{L^2}{V_0\overline{V}} + \rho T}{\sigma\sqrt{T}} + \mu_X^m\sqrt{T}\right)\end{aligned} \tag{3.6}$$

with α as defined in equation (3.5). If $\overline{V} < l_T$, set $\overline{V} = l_T$ when calculating this probability.

Proof. Using Lemma 3.2, the first probability in equation (3.3) can be derived as follows:

$$\mathbb{Q}^m\{X_T > {}_X\overline{V}\} = \mathbb{Q}^m\left\{\frac{X_T - X_0}{\sqrt{T}} > \frac{{}_X\overline{V} - X_0 - \mu_X^m T}{\sqrt{T}}\right\} \tag{3.7}$$

$$= N\left(\frac{\ln\frac{V_0}{\overline{V}} + \rho T}{\sigma\sqrt{T}} + \mu_X^m\sqrt{T}\right), \tag{3.8}$$

where the $(+\rho T)$-term in the argument of the normal distribution comes from the standardization of $\overline{V}$ (remember that ${}_X\overline{V} = \frac{1}{\sigma} \ln \frac{\overline{V}}{l_0 e^{\rho T}}$).

Similarly, for the second probability we get

$$\mathbb{Q}^m\{X_T > 2X_0 + {}_X\overline{V}\} = \mathbb{Q}^m \left\{ \frac{X_T - X_0}{\sqrt{T}} > \frac{X_0 + {}_X\overline{V} - \mu_X^m T}{\sqrt{T}} \right\} \tag{3.9}$$

$$= N\left(\frac{\ln \frac{l_0^2}{V_0 \overline{V}} + \rho T}{\sigma \sqrt{T}} + \mu_X^m \sqrt{T} \right). \tag{3.10}$$

□

Remark 3.2. Note that the extension from a constant to an exponential barrier changes μ_X^m. E.g., $\mu_X^B = \frac{r - \beta - \rho - \sigma^2/2}{\sigma}$ in the case of an exponential barrier collapses to the same expression with $\rho = 0$ in the case of a constant barrier.

A careful comparison of equations (3.6) and (3.4) reveals that a kind of "shortcut adaptation" of some formulas derived for the constant barrier case makes them applicable for the exponential barrier case. In the present case, we simply have to "adjust" α according to

$$\alpha_\rho = \frac{2(\mu_X^m - \rho/\sigma)}{\sigma}, \tag{3.11}$$

while still using equation (3.4) (instead of (3.6)). This holds because, as can be seen from the derivation and discussed above, the change in μ_X^m due to the exponential barrier is "undone" in the argument of the normal distributions by "adding back" $\rho\sqrt{T}/\sigma$.

Lemma 3.3 ("</down-and-out"). *Using the same definitions as in Lemma 3.2, the probability of* $\{V_T < \overline{V}, \tau \notin [0,T]\}$ *under various martingale measures* $\mathbb{Q}^m$ *is given by*

$$\begin{aligned} \mathbb{Q}^m\{V_T < \overline{V}, \tau \notin [0,T]\} = {} & \mathbb{Q}^m\{X_T > {}_X l_T, \tau \notin [0,T]\} \\ & - \mathbb{Q}^m\{X_T > {}_X\overline{V}, \tau \notin [0,T]\}. \end{aligned} \tag{3.12}$$

Proof. By complementarity,

$$\mathbb{Q}^m\{V_T < \overline{V}, \tau \notin [0,T]\} = \mathbb{Q}^m\{\tau \notin [0,T]\} - \mathbb{Q}^m\{V_T > \overline{V}, \tau \notin [0,T]\}.$$

Noting that $\mathbb{Q}^m\{\tau \notin [0,T]\} = \mathbb{Q}^m\{X_T > {}_X l_T, \tau \notin [0,T]\}$ completes the proof.

□

Given the ">/down-and-out"-probability, the derivation of the corresponding down-and-in-probability is straightforward:

Lemma 3.4 (">/down-and-in"). *Using the same definitions as in Lemma 3.2, the probability of* $\{V_T > \overline{V}, \tau \in [0,T]\}$ *under various martingale measures* $\mathbb{Q}^m$ *is given by*

$$\mathbb{Q}^m\{V_T > \overline{V}, \tau \in [0,T]\} = \mathbb{Q}^m\{V_T > \overline{V}\} - \mathbb{Q}^m\{V_T > \overline{V}, \tau \notin [0,T]\}. \quad (3.13)$$

Proof. Follows directly from the complementarity of the events $\{\tau \in [0,T]\}$ and $\{\tau \notin [0,T]\}$. □

Lemma 3.5 ("</down-and-in"). *Using the same definitions as in Lemma 3.2, the probability of* $\{V_T < \overline{V}, \tau \in [0,T]\}$ *under various martingale measures* $\mathbb{Q}^m$ *is given by*

$$\mathbb{Q}^m\{V_T < \overline{V}, \tau \in [0,T]\} = \mathbb{Q}^m\{V_T < \overline{V}\} - \mathbb{Q}^m\{V_T < \overline{V}, \tau \notin [0,T]\}. \quad (3.14)$$

Proof. Follows directly from the complementarity of the events $\{\tau \in [0,T]\}$ and $\{\tau \notin [0,T]\}$. □

We will also need the corresponding "up"-probabilities:

Lemma 3.6 ("</up-and-out"). *Using the same definitions as in Lemma 3.2, but assuming that* $V_0 < l_0$ *(so that* l_t *now represents an "up-barrier"), the probability of the event* $\{V_T < \overline{V}, \tau \notin [0,T]\}$ *under various martingale measures* $\mathbb{Q}^m$ *for* $\overline{V} \leq l_T$ *is given by*

$$\begin{aligned}\mathbb{Q}^m\{V_T < \overline{V}, \tau \notin [0,T]\} = {} & N\left(\frac{x\overline{V} - X_0 - \mu_X^m T}{\sqrt{T}}\right) \\ & - e^{-2\mu_X^m X_0} N\left(\frac{x\overline{V} + X_0 - \mu_X^m T}{\sqrt{T}}\right).\end{aligned} \quad (3.15)$$

If $\overline{V} > l_T$*, set* $\overline{V} = l_T$ *when calculating the probability.*

Proof. The proof proceeds along the lines of the proof of Lemma 3.2. We will only state some intermediate results. The $\mathbb{Q}^X$-probability is given by

$$\mathbb{Q}^X\{X_T < x\overline{V}, \tau \notin [0,T]\} = N\left(\frac{x\overline{V} - X_0}{\sqrt{T}}\right) - N\left(\frac{x\overline{V} + X_0}{\sqrt{T}}\right).$$

The densities are given by

$$\mathbb{Q}^X\{X_T \in dw_T^X, \tau \notin [0,T]\} = f(0, \sqrt{T})dw_T^X - f(-2X_0, \sqrt{T})dw_T^X.$$

Integration and square completion gives

$$\begin{aligned}\mathbb{Q}^m\{V_T < \overline{V}, \tau \notin [0,T]\} &= \int_{-\infty}^{x\overline{V}-X_0} f(\mu_X^m T, \sqrt{T}) dw_T^X \\ &\quad - e^{-2\mu_X^m X_0} \int_{-\infty}^{x\overline{V}-X_0} f(-2X_0 + \mu_X^m T, \sqrt{T}) dw_T^X \\ &= N\left(\frac{x\overline{V} - X_0 - \mu_X^m T}{\sqrt{T}}\right) \\ &\quad - e^{-2\mu_X^m X_0} N\left(\frac{x\overline{V} + X_0 - \mu_X^m T}{\sqrt{T}}\right).\end{aligned}$$

□

Lemma 3.7 (">/up-and-out"). *Using the same definitions as in Lemma 3.2, but assuming that* $V_0 < l_0$ *(so that* l_t *now represents an "up-barrier"), the probability of the event* $\{V_T > \overline{V}, \tau \notin [0,T]\}$ *under various martingale measures* $\mathbb{Q}^m$ *for* $\overline{V} < l_T$ *is given by*

$$\begin{aligned}\mathbb{Q}^m\{V_T > \overline{V}, \tau \notin [0,T]\} =& \mathbb{Q}^m\{V_T < l_T, \tau \notin [0,T]\} \\ &- \mathbb{Q}^m\{V_T < \overline{V}, \tau \notin [0,T]\},\end{aligned} \tag{3.16}$$

where the components are given by Lemma 3.6. If $\overline{V} \geq l_T$*, this probability is 0 (values above* $\overline{V}$ *cannot be reached without crossing the barrier).*

Proof. By complementarity,

$$\mathbb{Q}^m\{V_T > \overline{V}, \tau \notin [0,T]\} = \mathbb{Q}^m\{\tau \notin [0,T]\} - \mathbb{Q}^m\{V_T < \overline{V}, \tau \notin [0,T]\}.$$

Noting that $\mathbb{Q}^m\{\tau \notin [0,T]\} = \mathbb{Q}^m\{V_T < l_T, \tau \notin [0,T]\}$, equation (3.16) follows. □

Lemma 3.8 ("</up-and-in"). *Using the same definitions as in Lemma 3.2, but assuming that* $V_0 < l_0$ *(so that* l_t *now represents an "up-barrier"), the probability of the event* $\{V_T > \overline{V}, \tau \in [0,T]\}$ *under various martingale measures* $\mathbb{Q}^m$ *is given by*

$$\mathbb{Q}^m\{V_T < \overline{V}, \tau \in [0,T]\} = \mathbb{Q}^m\{V_T < \overline{V}\} - \mathbb{Q}^m\{V_T < \overline{V}, \tau \notin [0,T]\}, \tag{3.17}$$

where the components are given by Lemma 3.6.

Proof. Follows by complementarity. □

Lemma 3.9 (">/up-and-in"). *Using the same definitions as in Lemma 3.2, but assuming that $V_0 < l_0$ (so that l_t now represents an "up-barrier"), the probability of the event $\{V_T > \overline{V}, \tau \in [0,T]\}$ under various martingale measures $\mathbb{Q}^m$ is given by*

$$\mathbb{Q}^m\{V_T > \overline{V}, \tau \in [0,T]\} = \mathbb{Q}^m\{V_T > \overline{V}\} - \mathbb{Q}^m\{V_T > \overline{V}, \tau \notin [0,T]\}, \quad (3.18)$$

where the components are given by Lemma 3.7.

Proof. Follows by complementarity. □

The corresponding probabilities for constant and exponential barriers in terms of cumulative standard normal densities with the original variables as arguments for the cases covered in Lemmata 3.3 to 3.9 can be derived along the lines of Corollaries 3.1 and 3.2.

Another well-known result which will be needed later is stated in the following Lemma:

Lemma 3.10. *Given the stochastic process (X_t) under $\mathbb{Q}^m$:*

$$dX_t = \mu_X^m dt + dW_t^m, \quad X_0 = \underline{X},$$

its first passage time density to zero at s is given by

$$f^m(X_0, s) = \frac{X_0}{\sqrt{2\pi s^3}} \exp\left(-\frac{1}{2}\left(\frac{X_0 + \mu_X^m s}{\sqrt{s}}\right)^2\right).$$

3.3 Applications: Barrier Heavisides, Calls and Puts

To illustrate our pricing approach and for later reference, we provide here the pricing formulas for barrier heavisides and calls in the Black–Scholes model.

3.3.1 Barrier Heavisides

The pricing formula for the unrestricted (non-barrier) European heaviside in the Black–Scholes model was given in equation (3.1) (we restate it here for the case of an underlying paying continuous dividends at rate β):

$$H(S, K, T) = e^{-rT} N(d_2) \quad (3.19)$$

with

$$d_2(\cdot) = \frac{\ln(S_0/K) + (r - \beta + \sigma^2/2)T}{\sigma\sqrt{T}} - \sigma\sqrt{T}. \tag{3.20}$$

Lemma 3.11. *The values of the four barrier heavisides are given as follows:*

$$H_{cd}(S, K, T, l_0, \rho) = \begin{cases} e^{-rT}\mathbb{Q}^B\{S_T > K, \tau \notin [0, T]\} & \textit{if } d = O \\ e^{-rT}\mathbb{Q}^B\{S_T > K, \tau \in [0, T]\} & \textit{if } d = I \end{cases}, \tag{3.21}$$

with $c \in \{u, l\}$, $d \in \{I, O\}$, *and* $\mathbb{Q}^B\{\cdot\}$ *is given by Lemmata 3.2, 3.4, 3.7 and 3.9, respectively.* $c = u$ *denotes an up-barrier,* $c = l$ *a down-barrier,* $d = I$ *an "in-contract", and* $d = O$ *an "out-contract".*

Proof. From the definition of the heaviside and standard martingale valuation theory, we have

$$H_{cO}(S, K, T, l_0, \rho) = e^{-rT}\mathbb{E}^B[S_T > K, \tau \notin [0, T]]$$

and

$$H_{cI}(S, K, T, l_0, \rho) = e^{-rT}\mathbb{E}^B[S_T > K, \tau \in [0, T]].$$

□

If we had followed the standard approach, we would have ended up with four rather lengthy formulae, each with the need to distinguish between two cases ($K \geq l_T$ and $K < l_T$, respectively).

3.3.2 Barrier Calls

The pricing formula for the unrestricted (non-barrier) European call in the Black–Scholes model was given in Proposition 1.4 (we restate it here for convenience for the case of a dividend-paying underlying):

$$C(S, K, T) = e^{-\beta T}SN(d_1) - e^{-rT}KN(d_2) \tag{3.22}$$

with

$$d_1(\cdot) = \frac{\ln(S_0/K) + (r - \beta + \sigma^2/2)T}{\sigma\sqrt{T}} \tag{3.23}$$

and

$$d_2(\cdot) = d_1(\cdot) - \sigma\sqrt{T}. \tag{3.24}$$

Lemma 3.12. *The values of the four barrier calls are given as follows:*

$$C_{cd}(S, K, T, l_0, \rho) = e^{-\beta T} S \mathbb{Q}^S\{A_T\} - e^{-rT} K \mathbb{Q}^B\{A_T\}, \tag{3.25}$$

where

$$A_T = \begin{cases} \{S_T > K, \tau \notin [0,T]\} & \text{if } d = O \\ \{S_T > K, \tau \in [0,T]\} & \text{if } d = I \end{cases},$$

$c \in \{u, l\}$, $d \in \{I, O\}$, *and* $\mathbb{Q}^m\{\cdot\}$ ($m \in \{B, S\}$) *is given by Lemmata 3.2, 3.4, 3.7 and 3.9, respectively.* $c = u$ *denotes an up-barrier,* $c = l$ *a down-barrier,* $d = I$ *an "in-contract", and* $d = O$ *an "out-contract".*

Proof.

$$\begin{aligned} C_{cd}(S, K, T, l_0, \rho) &= e^{-rT} \mathbb{E}^B[(S_T - K) I_{\{A_T\}}] \\ &= e^{-rT} \left(\mathbb{E}^B[S_T] \mathbb{Q}^S\{A_T\} - K \mathbb{Q}^B\{A_T\} \right) \\ &= e^{-\beta T} S_0 \mathbb{Q}^S\{A_T\} - e^{-rT} K \mathbb{Q}^B\{A_T\}. \end{aligned}$$

□

Again, the standard approach would have left us with 4 formulae consisting of up to 6 different (non-barrier) options, each distinguishing between two cases.

Remark 3.3. Note that to emphasize that we are dealing with a constant barrier, we will use upper-case barrier symbols (i.e., $c \in \{U, L\}$ instead of $c \in \{u, l\}$) and drop the fifth argument (ρ). E.g., $C_{LO}(S, K, T, L)$ will denote a down-and-out call with a constant barrier.

3.3.3 Barrier Puts

Lemma 3.13. *The values of the four barrier puts are given as follows:*

$$P_{cd}(S, K, T, l_0, \rho) = e^{-rT} K \mathbb{Q}^B\{A_T\} - e^{-\beta T} S \mathbb{Q}^S\{A_T\}, \tag{3.26}$$

where

$$A_T = \begin{cases} \{S_T < K, \tau \notin [0,T]\} & \text{if } d = O \\ \{S_T < K, \tau \in [0,T]\} & \text{if } d = I \end{cases},$$

$c \in \{u, l\}$, $d \in \{I, O\}$, *and* $\mathbb{Q}^m\{\cdot\}$ ($m \in \{B, S\}$) *is given by Lemmata 3.3, 3.5, 3.6 and 3.8, respectively.* $c = u$ *denotes an up-barrier,* $c = l$ *a down-barrier,* $d = I$ *an "in-contract", and* $d = O$ *an "out-contract".*

Proof.

$$\begin{aligned} P_{cd}(S, K, T, L) &= e^{-rT}\mathbb{E}^B[(K - S_T)I_{\{A_T\}}] \\ &= e^{-rT}\left(K\mathbb{Q}^B\{A_T\} - \mathbb{E}^B[S_T]\mathbb{Q}^S\{A_T\}\right) \\ &= e^{-rT}K\mathbb{Q}^B\{A_T\} - e^{-\beta T}S_0\mathbb{Q}^S\{A_T\}. \end{aligned}$$

□

3.3.4 Relation to the Standard Approach

In the sequel, we will show how our approach taken in the previous section relates to the "standard" approach in barrier option pricing used, e.g., in Björk (1998, chapter 13).

We will use a down-and-in *call* option on a dividend-paying stock as a specific example and contrast pricing of this contract using our approach with valuation via the standard approach. For simplicity of exposition, we assume that the barrier is constant. The underlying follows a geometric Brownian motion under $\mathbb{P}$:

$$dS_t = (\mu - \beta)S_t dt + \sigma S_t dW_t,$$

where β denotes the dividend payout rate. Björk (1998, p. 192) provides the following (general) theorem for the pricing of down-and-in options (the notation has been adapted):

Theorem 3.1 (Down-and-in options, Björk (1998)). *Given a general claim with underlying* (S_t), *strike* K, *time* T *payoff* $\Gamma(S_T)$ *and time* $t = 0$ *price* $\Psi(\cdot, \Gamma(S_T))$, *the price of its down-and-in counterpart with constant barrier* L *(for* $S > L$*) is given by*

$$\Psi_{LI}(S, K, T, \Gamma, L) = \Psi(S, K, T, \Gamma^L) + \left(\frac{L}{S}\right)^{\alpha} \Psi\left(\frac{L^2}{S}, K, T, \Gamma_L\right), \qquad (3.27)$$

where $\Gamma_L(x) = \Gamma(x) \cdot I_{\{x>L\}}$, $\Gamma^L = \Gamma(x) \cdot I_{\{x \leq L\}}$, *and* α *is defined as in equation* (3.5).

To apply this theorem, we first have to find the conditional payoff functions Γ_L and Γ^L for the two cases $L < K$ and $L \geq K$. Figure 3.2 provides a graph visualizing the four results.

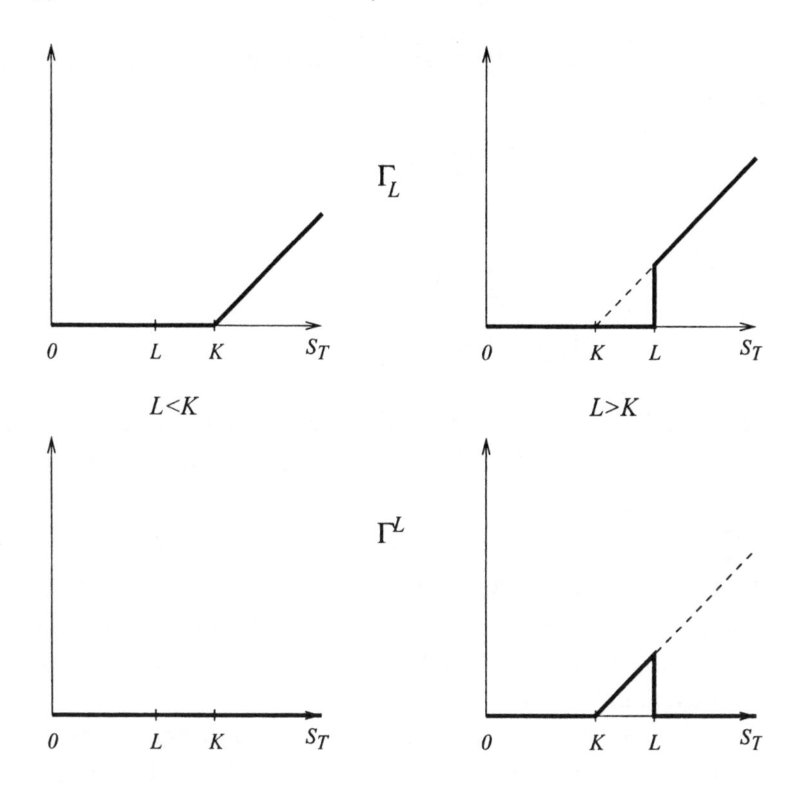

Figure 3.2: Conditional call payoffs for barriers above/below the strike

For $L < K$, we get

$$\Gamma_L = (S - K)_+ = \Gamma(C(S, K, T))$$

and

$$\Gamma^L = 0,$$

and for $L \geq K$, we have

$$\Gamma_L = (S - K)I_{\{S>L\}} = \Gamma(C(S, L, T)) + (L - K)\Gamma(H(S, L, T))$$

and

$$\Gamma^L = (S - K)_+ I_{\{S<L\}} = \Gamma(C(S, K, T)) - \Gamma(C(S, L, T)) \\ - (L - K)\Gamma(H(S, L, T))$$

Therefore, the value of the down-and-in call according to Björk (1998) is given as follows:

$$C_{LI}(S,K,T,L) = \begin{cases} \left(\frac{L}{S}\right)^{\alpha} C\left(\frac{L^2}{S},K,T\right) & L < K \\ C(S,K,T) - C(S,L,T) - (L-K)H(S,L,T) & \\ + \left(\frac{L}{S}\right)^{\alpha} \left(C\left(\frac{L^2}{S},L,T\right)\right. & L \geq K \\ \left. -(L-K)H\left(\frac{L^2}{S},L,T\right)\right) & \end{cases}$$

We compare this to the result we obtained using the more direct probabilistic approach (cf. equation (3.25), restated here for convenience):

$$C_{LI}(S,K,T,L) = e^{-\beta T} S_0 \mathbb{Q}^S\{A_T\} - e^{-rT} K \mathbb{Q}^B\{A_T\},$$

where the probabilities for the event $A_T = \{V_T > K, \tau \in [0,T]$ under measures $\mathbb{Q}^B$ and $\mathbb{Q}^S$ are given by Lemma 3.4. It can be clearly seen that, once the corresponding probabilities are known, our approach is more direct, and more convenient to use.

In general, given the price of the up/down-and-in claim, pricing the corresponding up/down-and-out claim becomes straightforward via the so-called *in-out-parity*:

$$\Psi_{\cdot I}(\cdot) = \Psi(\cdot) - \Psi_{\cdot O}(\cdot). \tag{3.28}$$

This relation holds for options knocked in or out at barriers below the current asset price (down-and-in/out options) as well as those knocked in/out at barriers above the current asset price (up-and-in/out options). It is just a corollary of the proofs of the corresponding Lemmata stating the in- and out-probabilities.

3.4 Numerical Examples

Example 3.1. Assume that the current stock price is $S = 100$, the strike $K = 110$, the risk-free interest rate $r = 0.05$, continuous dividend yield $\beta = 0.04$, stock return volatility $\sigma = 0.3$ and the expiration time $T = 2$ (years). We want to calculate the value of an unrestricted European heaviside option $H(S,K,T)$ and that of an unrestricted European call option using these data.

For the heaviside, we get (using equation (3.19)):

$$H(100,110,2) = e^{-0.05\cdot 2}N(d_2) = 0.3152 \tag{3.29}$$

with

$$d_2 = \frac{\ln(100/110) + 2(0.05 - 0.04 + 0.3^2/2)}{0.3\sqrt{2}} - 0.3\sqrt{2}. \tag{3.30}$$

The call price is (using equation (3.22)):

$$\begin{aligned} C(100,110,2) &= 100e^{-0.04\cdot 2}N(d_2 + 0.3\sqrt{2}) - 110e^{-0.05\cdot 2}N(d_2) \\ &= 12.7535 \end{aligned} \tag{3.31}$$

with d_2 as defined above for the heaviside.

Example 3.2. Same data as in Example 3.1 plus a *constant* barrier $L = 80$. We want to calculate the values of the down-and-out counterparts of the options in Example 3.1.

For the heaviside, we start by remembering from equation (1.25) that

$$\mu_X^B = \frac{0.05 - 0.04 - 0.3^2/2}{0.3} = -0.1167.$$

Then, we use equations (3.21) and (3.4) to get

$$\begin{aligned} H_{LO}(100,110,2,80) &= e^{-rT}\mathbb{Q}^B\{S_T > K, \tau \notin [0,T]\} \\ &= e^{(-2\cdot 0.05)}\left[N\left(\frac{\ln\left(\frac{100}{110}\right)}{0.3\sqrt{2}} - 0.1167\sqrt{2}\right)\right. \\ &\quad \left. - \left(\frac{80}{100}\right)^{-\frac{2\cdot 0.1167}{0.3}} N\left(\frac{\ln\left(\frac{80^2}{100\cdot 110}\right)}{0.3\sqrt{2}} - 0.1167\sqrt{2}\right)\right] \\ &= 0.2348. \end{aligned} \tag{3.32}$$

For the down-and-out call, we start by remembering from equation (1.23) that

$$\mu_X^S = -0.1167 + 0.3 = 0.1833.$$

Then, we use equations (3.25) and (3.4) to get

$$\begin{aligned}
C_{LO}(100,110,2,80) &= e^{-\beta T} S_0 \mathbb{Q}^S\{S_T > K, \tau \notin [0,T]\} \\
&\quad - e^{-rT} K \mathbb{Q}^B\{S_T > K, \tau \notin [0,T]\} \\
&= 100 e^{(-2\cdot 0.04)} \left[N\left(\frac{\ln\left(\frac{100}{110}\right)}{0.3\sqrt{2}} + 0.1833\sqrt{2} \right) \right. \\
&\quad \left. - \left(\frac{80}{100}\right)^{\frac{2\cdot 0.1833}{0.3}} N\left(\frac{\ln\left(\frac{80^2}{100\cdot 110}\right)}{0.3\sqrt{2}} + 0.1833\sqrt{2} \right) \right] \\
&\quad - 110 e^{(-2\cdot 0.05)} \left[N\left(\frac{\ln\left(\frac{100}{110}\right)}{0.3\sqrt{2}} - 0.1167\sqrt{2} \right) \right. \\
&\quad \left. - \left(\frac{80}{100}\right)^{-\frac{2\cdot 0.1167}{0.3}} N\left(\frac{\ln\left(\frac{80^2}{100\cdot 110}\right)}{0.3\sqrt{2}} - 0.1167\sqrt{2} \right) \right] \\
&= 10.7411.
\end{aligned} \tag{3.33}$$

Comparing the results of the down-and-out options in this example to the prices of their non-barrier counterparts calculated in Example 3.1, we see that the additional requirement of not hitting the barrier during the options' lifetimes leads to a decrease in the options' prices.

Example 3.3. Same data as in Example 3.2, but now we consider an exponential barrier with $l_0 = 80$ and $\rho = 0.06$. We want to calculate the prices of the down-and-out heaviside and call from Example 3.2 under these assumptions.

We start by remembering from the proof of Lemma 3.2 and Remark 3.2 that the assumption of an exponential barrier changes the drifts μ_X^m. The new drifts are given by

$$\mu_X^B = \frac{0.05 - 0.04 - 0.06 - 0.3^2/2}{0.3} = -0.3167$$

and

$$\mu_X^S = -0.3167 + 0.3 = -0.0167.$$

For the heaviside, we use equations (3.21) and (3.6) to get

$$
\begin{aligned}
H_{LO}(\cdot) &= e^{-rT}\mathbb{Q}^B\{S_T > K, \tau \notin [0,T]\} \\
&= e^{(-2\cdot 0.05)}\left[N\left(\frac{\ln\left(\frac{100}{110}+2\cdot 0.06\right)}{0.3\sqrt{2}} - 0.3167\sqrt{2}\right)\right. \\
&\quad \left. - \left(\frac{80}{100}\right)^{-\frac{2\cdot 0.3167}{0.3}} N\left(\frac{\ln\left(\frac{80^2}{100\cdot 110}+2\cdot 0.06\right)}{0.3\sqrt{2}} - 0.3167\sqrt{2}\right)\right] \\
&= 0.2070.
\end{aligned}
\tag{3.34}
$$

For the call, we use equations (3.25) and (3.6) to get

$$
\begin{aligned}
C_{LO}(\cdot) &= e^{-\beta T}S_0\mathbb{Q}^S\{S_T > K, \tau \notin [0,T]\} \\
&\quad - e^{-rT}K\mathbb{Q}^B\{S_T > K, \tau \notin [0,T]\} \\
&= 100e^{(-2\cdot 0.04)}\left[N\left(\frac{\ln\left(\frac{100}{110}+2\cdot 0.06\right)}{0.3\sqrt{2}} - 0.0167\sqrt{2}\right)\right. \\
&\quad \left. - \left(\frac{80}{100}\right)^{-\frac{2\cdot 0.0167}{0.3}} N\left(\frac{\ln\left(\frac{80^2}{100\cdot 110}+2\cdot 0.06\right)}{0.3\sqrt{2}} - 0.0167\sqrt{2}\right)\right] \\
&\quad - 110e^{(-2\cdot 0.05)}\left[N\left(\frac{\ln\left(\frac{100}{110}+2\cdot 0.06\right)}{0.3\sqrt{2}} - 0.3167\sqrt{2}\right)\right. \\
&\quad \left. - \left(\frac{80}{100}\right)^{-\frac{2\cdot 0.3167}{0.3}} N\left(\frac{\ln\left(\frac{80^2}{100\cdot 110}+2\cdot 0.06\right)}{0.3\sqrt{2}} - 0.3167\sqrt{2}\right)\right] \\
&= 10.0438.
\end{aligned}
\tag{3.35}
$$

Comparing equations (3.34) and (3.35) to equations (3.31) and (3.33), we find that the prices of the down-and-out options in the case of an exponentially increasing barrier are smaller than for the constant barrier case (when the barriers are equal at time $t = 0$). This is to be expected: Since the barrier "moves towards" the underlying, it becomes more likely to be reached, increasing the chance of the option being "knocked out". A lower chance of survival leads to a smaller expected payoff for the option.

Chapter 4

A Probabilistic, Firm Value Based Security Pricing Framework

In this chapter, we review a framework for pricing corporate securities in the presence of credit risk developped by Ericsson and Reneby (1998, 2001). We extend this framework considerably by deriving valuation formulae for an array of additional claims. Its flexibility will be demonstrated in Chapter 5, where we will show that many classical results can be very conveniently and elegantly derived within this framework.

4.1 Ericsson and Reneby (1998)

Ericsson and Reneby (1998) provide a highly flexible framework for the valuation of corporate securities. Since this model – together with the extension for the valuation of options on corporate securities described in Section 6.1 – is the basis for most of the material presented in the following chapters, we describe it here in more detail.

For the reader who wants to "go to the sources", it may be helpful to note that there are some differences in notation between Ericsson and Reneby (1996) and Ericsson and Reneby (1998), the obvious reason being that the latter paper (written earlier) has been revised for publication in 1998, whereas the former was not. The notation used there is largely based on a working paper version of Ericsson and Reneby (1998) published in Ericsson (1997).

The fundamental idea behind this framework is to decompose the payoffs of corporate securities into the payoffs of standardized building blocks. Once the values of the building blocks are known, the securities can then be valued as sums of the values of the building blocks. Whereas the framework presented here uses a probabilistic setting, there are also papers applying this building-block approach in a PDE setting (see, e.g., Goldstein, Ju, and Leland (2001) or Christensen, Flor, Lando, and Miltersen (2002)).

4.1.1 Assumptions

The main assumptions in Ericsson and Reneby (1998) are as follows:[1]

Assumption 4.1. *The variable determining the liquidation value of the firm's assets, ν_t, at some future date τ, follows a geometric Brownian motion:*

$$d\nu_t = \mu \nu_t dt + \sigma \nu_t dW_t,$$

with $\nu_0 = \underline{\nu}$ and (W_t) is a Wiener process under the objective probability measure.

Leland (1994, p. 1217) already notes that the (mathematically convenient, but economically implausible) assumption of continuously traded assets made in many early papers applying contingent claims analysis to the pricing of a firm's securities is unnecessary. If at least one of the firm's securities is traded (continuously), the (unobservable) asset value process can be constructed from the process of the traded security.[2] Ericsson and Reneby (1998) note that an additional assumption necessary for this argument to work is that the firm's assets will be traded at *some* time in the future, τ.

Assumption 4.2. *Capital markets are frictionless for at least some large investors, i.e. there are no transaction costs, assets are perfectly divisible, arbitrage opportunities are ruled out, there are no restrictions on short sales, and borrowing and lending takes place at the risk-free interest rate. At least one security on the firm's balance sheet is traded.*

Assumption 4.3. *The risk-free interest rate, r, is constant.*

Denoting by V_t the time t value of a claim with a payoff of ν_τ, this value at time $t = 0$ is given by

$$V_0 = e^{-r\tau} \mathbb{E}^B [\nu_\tau] = \nu_0 e^{(\mu + \beta - \lambda\sigma - r)\tau},$$

[1] For a detailed discussion of the assumptions, see Ericsson and Reneby (1998, pp. 145ff.).

[2] Taxes may be an obstacle to this replication, see e.g. Goldstein, Ju, and Leland (2001).

where $\mathbb{Q}^B$ denotes the martingale measure, λ is the market price of risk for the operations of the firm, and β is the asset payout rate. V_t will be called the *asset value* (or the *value of the corresponding all-equity firm*).

Assumption 4.4. *Default occurs if*

$$V_t \leq L, \text{ for some } t \leq T,$$

or

$$V_T < F.$$

This assumption means that default is triggered either if the value of assets falls below a prespecified (constant) boundary L during the lifetime of the securities with finite maturity (we denote the first passage time of (V_t) to L with τ), or if at maturity the promised payments F (e.g., repayment of principal in the case of debt) cannot be made. In case of default (or "reorganization", which seems to be the preferred term in the more recent literature), the value of the firm's assets is distributed to claimants.

Assumption 4.5. *The (possibly random) fractions of the value of assets distributed to claimants in case of reorganization are independent of time and the level of asset value.*

This assumption is only for convenience, since we do not want to focus on issues like debt renegotiation here. The fractions could, e.g., also be determined as the outcome of a game played between securityholders.

4.1.2 Valuation of the Building Blocks

The following securities are used as building blocks, or "basic claims": the down-and-out heaviside, the down-and-out call, the unit down-and-in claim, and the down-and-out asset claim (the latter two will be defined shortly). For most of these claims, the probabilities of the event $A_T = \{\tau \nleq T, V_T > F\}$ under different probability measures play a central role. Therefore, these probabilities are defined in Lemma 4.1 before we give the pricing formulae for the building blocks.

Lemma 4.1. *The probabilities of the event* $A_T = \{\tau \notin [0,T], V_T > F\}$ *under the probability measures* $\mathbb{Q}^m (m \in \{B, V, G\})$ *are*

$$\mathbb{Q}^m\{A_T\} = N\left(d_T^m\left(\frac{V_0}{F}\right)\right) - \left(\frac{V_0}{L}\right)^{-\frac{2}{\sigma}\cdot\mu_X^m} N\left(d_T^m\left(\frac{L^2}{V_0 \cdot F}\right)\right) \tag{4.1}$$

where

$$d_t^m(x) = \frac{\ln x}{\sigma\sqrt{t}} + \mu_X^m \cdot \sqrt{t}, \tag{4.2}$$

$$\begin{cases} \mu_X^B = \frac{r-\beta-\sigma^2/2}{\sigma} \\ \mu_X^V = \mu_X^B + \sigma \\ \mu_X^G = \mu_X^B - \theta\sigma \end{cases} \tag{4.3}$$

and

$$\theta = \frac{\sqrt{(\mu_X^B)^2 + 2r} + \mu_X^B}{\sigma} \tag{4.4}$$

If $F < L$*, set* $F = L$ *in the above expression.*

The values of these basic building blocks (the measure $\mathbb{Q}^G$ uses the unit down-and-in claim as numeraire, see Definition 4.1 below) are given as follows (the formulae for the down-and-out call and heaviside have already been given in Section 3.3 and are restated here for convenience):

Lemma 4.2. *The time* $t = 0$ *price of a down-and-out heaviside on* V *with strike* F*, maturity* T *and barrier* L *is given by*

$$H_{LO}(V, F, T, L) = e^{-rT} \cdot \mathbb{Q}^B\{A\}.$$

Lemma 4.3. *The time* $t = 0$ *price of a down-and-out call on* V *with strike* F*, maturity* T *and barrier* L *is given by*

$$C_{LO}(V, F, T, L) = Ve^{-\beta T} \cdot \mathbb{Q}^V\{A\} - e^{-rT}F \cdot \mathbb{Q}^B\{A\}.$$

Definition 4.1. A *unit down-and-in claim* $G_{LI}(V, T, L)$ is defined via the following payoff function:

$$\Gamma_\tau(G_{LI}(V, T, L)) = I_{\{\tau \leq T\}}.$$

The payoff function of this claim is depicted graphically in Figure 4.1. In a sense, this claim pays off like a heaviside, but at an uncertain future time.

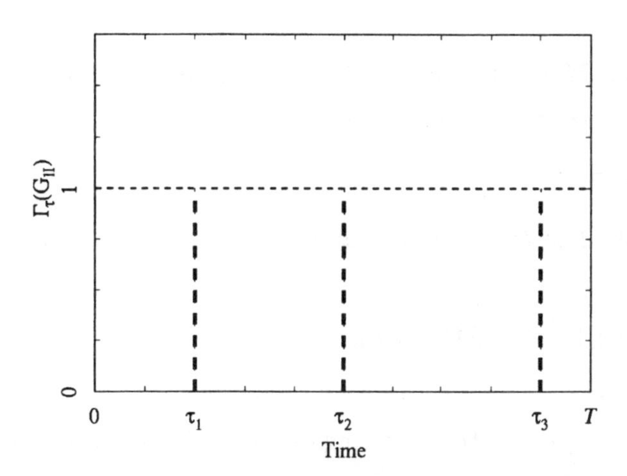

Figure 4.1: Possible payoffs of a unit down-and-in claim for various bankruptcy times τ_i

Lemma 4.4. *The time* $t = 0$ *price of a unit down-and-in claim with maturity* T *and barrier* L *is given by*

$$G_{LI}(V, T, L) = G_{LI}(V, \infty, L) \cdot (1 - \mathbb{Q}^G\{\tau \nleq T, V_T > L\})$$

with

$$G_{LI}(V, \infty, L) = \left(\frac{V}{L}\right)^{-\theta}.$$

Proof. The proof works via integration of the first passage time density given in Lemma 3.10, see Ericsson and Reneby (1998, p. 161). □

For ease of interpretation, the notion of an *asset claim* (although formally redundant) is also useful:

Definition 4.2. An *asset claim* $\Omega(V, T)$ is defined via the following payoff function:

$$\Gamma_T(\Omega(V, T)) = V_T.$$

Obviously, the asset claim is just a call option on V with a strike price of 0.

Lemma 4.5. *The time* $t = 0$ *value of a claim with payoff* V_T *at time* $t = T$ *("asset claim") is given by*

$$\Omega(V, T) = Ve^{-\beta T}. \tag{4.5}$$

Lemma 4.6. *The time* $t = 0$ *value of a down-and-out asset claim is given by*

$$\Omega_{LO}(V, T, L) = C_{LO}(V, 0, T, L). \tag{4.6}$$

Proofs. Omitted. □

In addition, we state the following results for later reference (cf. Ericsson and Reneby (1998, p. 151)).

Definition 4.3. A *down-and-out unit stream* $U_{LO}(V, T, L)$ is defined via the following payoff rate function:

$$\Gamma_t(U_{LO}(V, T, L)) = \Gamma_t(H_{LO}(V, 0, t, L)) \quad t \in [0, T].$$

The payoff rate of this claim is depicted graphically in Figure 4.2. Before default occurs, the payoff rate of the claim is 1. At time τ, the payoff rate drops to zero and stays there.

Lemma 4.7. *The time* $t = 0$ *value of a unit stream with maturity* T*, down-and-out at a barrier* L*, is given by*

$$U_{LO}(V, T, L) = \frac{1}{r}(1 - G_{LI}(V, T, L) - H_{LO}(V, 0, T, L)). \tag{4.7}$$

For infinite maturity, this expression simplifies to

$$U_{LO}(V, \infty, L) = \frac{1}{r}(1 - G_{LI}(V, \infty, L)). \tag{4.8}$$

Proof. See Ericsson and Reneby (1998, p. 161f.) or the proof of Lemma 4.12. □

Definition 4.4. A *down-and-out asset stream* $O_{LO}(V, T, L)$ is defined via the following payoff rate function:

$$\Gamma_t(O_{LO}(V, T, L)) = \Gamma_t(C_{LO}(V, 0, t, L)) \quad t \in [0, T].$$

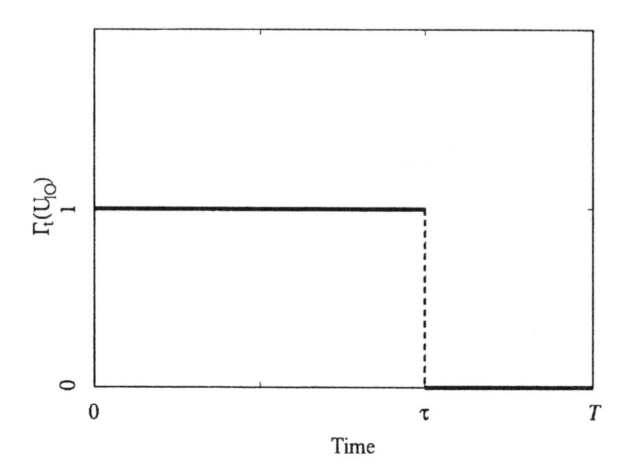

Figure 4.2: Payoff rate function of a down-and-out unit stream

The payoff rate of this claim is depicted graphically in Figure 4.3. Before default, the payoff rate of the claim is V_t. At time τ, the payoff rate drops to zero and stays there.

Lemma 4.8. *The time* $t = 0$ *value of an asset stream, down-and-out at a barrier* L, *is given by*

$$O_{LO}(V, T, L) = \frac{1}{\beta}(V - L \cdot G_{LI}(V, T, L) - \Omega_{LO}(V, T, L)). \tag{4.9}$$

For infinite maturity, this expression simplifies to

$$O_{LO}(V, \infty, L) = \frac{1}{\beta}(V - L \cdot G_{LI}(V, \infty, L)). \tag{4.10}$$

Proof. See Ericsson and Reneby (1998, p. 162f.) or the proof of Lemma 4.13. □

4.1.3 Results

The central proposition in Ericsson and Reneby (1998) is the following:

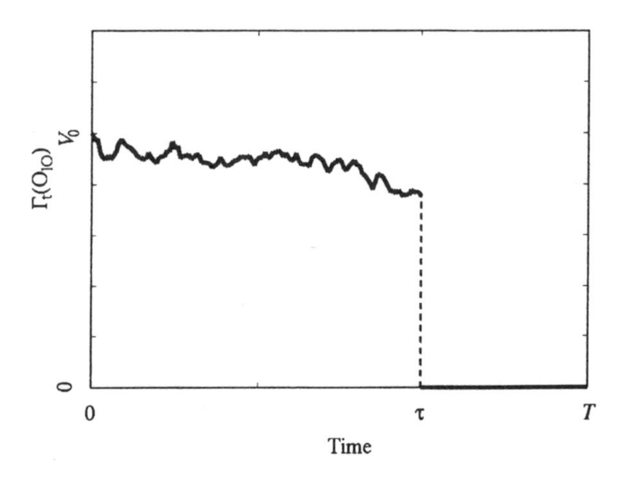

Figure 4.3: Payoff rate function of a down-and-out asset stream

Proposition 4.1. *A corporate security* CS *with contracted payments*

$$\Gamma(CS(\cdot)) = a\Gamma(\Omega) + \sum_i b^{(i)}\Gamma(C^{(i)}(\cdot)) + \sum_i c^{(i)}\Gamma(H^{(i)}(\cdot))$$

for $t \leq \tau$ *can be valued as*

$$CS(\cdot) = a\Omega_{LO} + \sum_i b^{(i)}C^{(i)}_{LO}(\cdot) + \sum_i c^{(i)}H^{(i)}_{LO}(\cdot) + \varphi^{CS}(L-k)G_{LI}(\cdot).$$

for $t \leq \tau$*, where* a, b, c *denote the numbers (fractions) of the respective claims, and* φ^{CS} *denotes the fraction of* V_τ *received by the holders of security* CS*. The summation operator should be understood to encompass integrals when applicable.*

If the payoffs of a security can be decomposed into the payoffs of other, more basic securities (calls, heavisides,...), the price of the security can be determined as the sum of the prices of the basic securities. In principle, the applicability of this framework is not limited to a (firm value based) Black–Scholes world, but is convenient for all models in which closed-form expressions for the prices of the basic building blocks can be derived.

As an example, Ericsson and Reneby (1998, pp. 152 ff.) derive the values of equity and debt if the only debt outstanding is a coupon bond. They do so by decomposing the payoffs to the securities into

- payoffs at maturity (conditional on $\tau \not< T$),
- intermediate payoffs in solvency, and
- payoffs in case of default (or "reorganization", as they prefer to call the first passage time of the firm value to the boundary).

To illustrate their approach, we restate part of this example here (with slight changes in notation): the derivation of the value of equity.

The bond has a face value of D and pays constant coupons C at times $t_i : i+1, \ldots, N$. Coupons are tax deductible (whereas repayment of principal is not), and the corporate tax rate is ζ. Ericsson and Reneby (1998) allow for violations of the absolute priority rule. Denoting by φ^{CS} (with $CS \in \{E, D\}$) the fraction of asset value, net of reorganization costs k, paid out to the holders of the corresponding corporate securities, this means that $\varphi^E > 0$. To facilitate the exposition, they confine themselves to the case $D \geq L \geq k$.

The payoffs received by equity are decomposed as discussed above into three components:

$$E = E^M + E^S + E^\tau,$$

where E^M denotes the time $t = 0$ value of the payoff to equity at maturity, E^S the value of the payoffs to equity prior to maturity conditional on no prior default, and E^τ the value of the payoff to equity in case of default (or reorganization) prior to maturity.

At maturity (time $t = T$), the payoff to equity (conditional on the barrier not having been hit before) is given by

$$\Gamma(E^M) = \begin{cases} V_T - D & \text{if } V_T \geq D \\ \varphi^E(V_T - k) & \text{if } D > V_T \geq L \end{cases}. \tag{4.11}$$

The first line in equation (4.11) is the payoff of one call on V with strike D, and the second line is the payoff of φ^E calls on V with strike k.[3] Thus,

[3]Assuming a diffusion process for (V_t), note that $V_T < L$ is not possible if $\tau \not< T$ and $V_0 > L$.

we can rewrite equation (4.11) as follows:

$$\begin{aligned}\Gamma(E^M) &= \Gamma(C(V,D,T)) + \varphi^E \Gamma(C(V,k,T)) \cdot I_{\{V_T<D\}} \\ &= \Gamma(C(V,D,T)) + \varphi^E \Gamma(C(V,k,T)) \\ &\quad - \varphi^E \Gamma(C(V,D,T)) - \varphi^E (D-k) \Gamma(H(V,D,T)).\end{aligned} \tag{4.12}$$

Therefore, using equation (4.12) and Proposition 4.1, the time $t = 0$ value of the "maturity payoff portion" of equity is given by

$$E^M = \varphi^E C_{LO}(V,k,T) + (1-\varphi^E) C_{LO}(V,D,T) - \varphi^E (D-k) H_{LO}(V,D,T).$$

Coupon payments decrease the value of equity. Each coupon payment represents a payoff of C at the due date of the coupon, t_i. Coupons are paid as long as $V_t > L$. The value of coupons (representing positive payoffs for debt, but negative payoffs for equity) is given by

$$D^S = C \sum_{i+1}^{N} H_{LO}(V,L,t_i).$$

However, since coupons are tax deductible, part of the coupons is financed by the treasury, and only a fraction $(1-\zeta)$ of the cost is borne by equity. Therefore, the value of coupons from the viewpoint of equity is given by

$$E^S = -(1-\zeta) C \sum_{i+1}^{N} H_{LO}(V,L,t_i).$$

In case of default prior to maturity because of the firm value hitting the barrier, the payoff to equity is given by

$$\Gamma(E^\tau) = \varphi^E (V_\tau - k) = \varphi^E (L-k), \tag{4.13}$$

where the second equation follows from the fact that we know the firm value in case of default. Using equation (4.13) and Proposition 4.1, the value of the payoff to equity in case of default prior to maturity is given by

$$E^\tau = \varphi^E (L-k) G_{LI}(V,L,T).$$

Summing up, we get the value of equity:

$$\begin{aligned}E(\cdot) &= \varphi^E C_{LO}(V,k,T) + (1-\varphi^E) C_{LO}(V,D,T) \\ &\quad - \varphi^E (D-k) H_{LO}(V,D,T) + \varphi^E (L-k) G_{LI}(V,L,T) \\ &\quad - (1-\zeta) C \sum_{i+1}^{N} H_{LO}(V,L,t_i).\end{aligned}$$

In the remainder of the paper, Ericsson and Reneby (1998) discuss possible extensions of their framework to allow for more complex capital structures and strategic debt service as in the model of Mella-Barral and Perraudin (1997).

At first sight, the fact that there are only a limited number of settings which lead to closed-form solutions may be viewed as a limitation of the approach presented here. However, even if numerical methods have to be used to actually *compute* prices, this approach is very powerful for deriving pricing equations.

4.2 Ericsson and Reneby (2001)

Ericsson and Reneby (2001) provide an extension to their 1998 work, which is motivated by the following observation: Since the value of a firm's assets is assumed to grow over time, the value of debt should also increase with time, otherwise the firm's debt-equity ratio would eventually go to zero (conditional on no prior default). While this is certainly true, it does not imply, however, that using a constant barrier as an approximation will lead to bad model fit, especially when it comes to pricing securities with short lifespans. Nevertheless, we will describe this extension in detail here, and examine its effect on option prices later in Chapter 9.

An important assumption in Ericsson and Reneby (2001) is that debt, although made up of many small debt issues, can be approximated at the aggregate level by a continuous growth rate ρ:

$$dD(t) = \rho D(t)dt,$$

where $D(t)$ denotes total debt principal outstanding at time t. Single debt issues can then be valued assuming that they are too small to have a significant effect on total debt. This implies, however, that single debt issues have no effect whatsoever on equity, whose value depends only on total debt.

In this environment with equity and total debt increasing exponentially, it makes sense to assume that the default barrier also increases exponentially with time:

$$dl_t = \rho l_t dt.$$

Similar to their 1998 work, the Ericsson and Reneby (2001) model features taxes and deviations from the absolute priority rule.

The first important extension is that of Lemma 4.1 (see p. 61) to the case of an exponential barrier. Whereas in Ericsson and Reneby (2001) a restricted version is given, Reneby (1998, p. 29) provides the general version.

Lemma 4.9. *The probabilities of the event* $A = \{V_T > F, \tau \nleq T\}$ *under the probability measures* $\mathbb{Q}^m (m \in \{B, V, G\})$ *are*

$$\mathbb{Q}^m\{A\} = N\left(d_T^m\left(\frac{V_0}{Fe^{-\rho T}}\right)\right) - \left(\frac{V_0}{l_0}\right)^{-\frac{2}{\sigma}\cdot\mu_X^m} N\left(d_T^m\left(\frac{l_0^2}{V_0 \cdot Fe^{-\rho T}}\right)\right) \tag{4.14}$$

where

$$d_t^m(x) = \frac{\ln x}{\sigma\sqrt{t}} + \mu_X^m \cdot \sqrt{t}, \tag{4.15}$$

$$\begin{cases} \mu_X^B = \frac{r-\beta-\rho-\sigma^2/2}{\sigma} \\ \mu_X^V = \mu_X^B + \sigma \\ \mu_X^G = \mu_X^B - \theta(r)\sigma \end{cases} \tag{4.16}$$

and

$$\theta(y) = \frac{\sqrt{(\mu_X^B)^2 + 2y} + \mu_X^B}{\sigma} \tag{4.17}$$

If $F < l_T$, *set* $F = l_T$ *in the above expression.*

Comparing Lemmata 4.1 and 4.9, carefully note the following: First, the drifts μ_X^m have been changed to account for the growth rate of the barrier. However, similar to our discussion in Remark 3.2 (p. 46), this change has to be "undone" for the arguments $d_t^m(\cdot)$ of the normal distributions (this was illustrated in Example 3.3). Second, the definition of θ has been extended to allow for arguments other then r (whereas only $\theta = \theta(r)$ was needed in the constant barrier case, and therefore a restricted version was provided in Lemma 4.1). The drifts μ_X^B in $\theta(\cdot)$ have also been modified accordingly.

In addition, they provide the following extension to the unit down-and-in claim described in Lemma 4.4 (a "dollar with interest down-and-in claim" with interest rate ρ, where the "interest rate" equals the growth rate of the barrier):

Definition 4.5. A *barrier-increasing unit down-and-in claim* $G^{\rho}_{lI}(\cdot)$ with infinite maturity is defined via the following payoff function:

$$\Gamma_{\tau}(G^{\rho}_{lI}(V, \infty, l_0, \rho)) = e^{\rho\tau} I_{\{\tau \leq T\}}.$$

Figure 4.4 provides a graphical illustration of the payoff function of this claim.

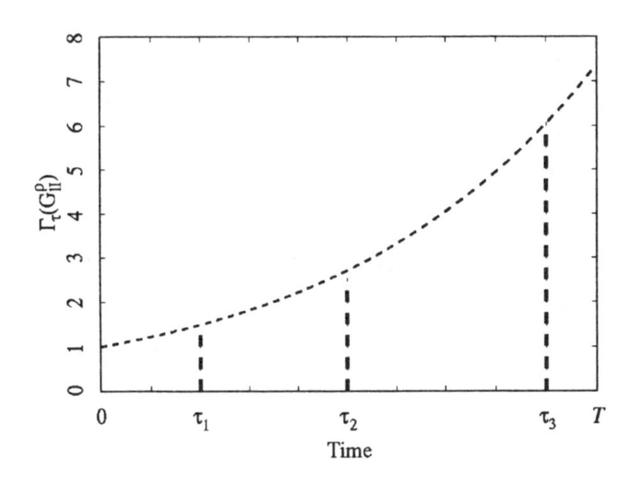

Figure 4.4: Possible payoffs of a barrier-increasing unit down-and-in claim for various bankruptcy times τ_i

Lemma 4.10. *The time* $t = 0$ *price of a barrier-increasing down-and-in claim is given by*

$$G^{\rho}_{lI}(V, \infty, l_0, \rho) = \left(\frac{V}{l_0}\right)^{-\theta(r-\rho)}$$

if $(\mu^B_X)^2 \geq -2(r - \rho)$.

Proof. See Ericsson and Reneby (2001, p. 44). □

Implicitly, they also derive the value of a barrier-increasing down-and-out unit stream with infinite maturity:

Definition 4.6. A *barrier-increasing down-and-out unit stream* $U^{\rho}_{lO}(\cdot)$ with infinite maturity is a security with the following payoff rate function:

$$\Gamma_t(U^{\rho}_{lO}(V,\infty,l_0,\rho)) = e^{\rho t}\Gamma_t(H_{lO}(V,0,t,l_0,\rho)) \quad t \geq 0.$$

Figure 4.5 illustrates the payoff rate of this claim.

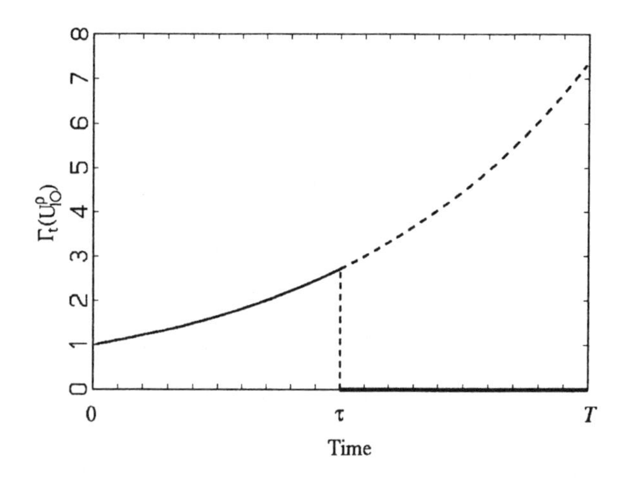

Figure 4.5: Payoff rate function of a barrier-increasing down-and-out unit stream

Lemma 4.11. *The price of the barrier-increasing down-and-out unit stream is (for* $r \neq \rho$*) given by*

$$U^{\rho}_{lO}(V,\infty,l_0,\rho) = \frac{1}{r-\rho}(1 - G^{\rho}_{lI}(V,\infty,l_0,\rho)). \tag{4.18}$$

For $r = \rho$*, the price is found as the limiting case.*

Proof. See, Ericsson and Reneby (2001, p. 45f.) (in the derivation of their Proposition 1). □

Remark 4.1. Note that $\rho > r$ in equation (4.18) does not lead to a negative value for the claim, because

$$\{\rho > r\} \Leftrightarrow \{G^{\rho}_{lI}(V,\infty,l_0,\rho) > 1\}.$$

Assuming that carrying forward of neither profits nor losses is admitted, Ericsson and Reneby (2001) note that the cash-flow available to equity at any given point in time is equal to

$$F_t^E = \beta V_t - (1-\zeta)C_t + d(V_t, t),$$

where $d(V_t, t)$ denotes cash inflow from new debt issues, and the remaining variables are defined as in the previous section. Total payoff to equity holders consists of the integral of this cash-flow over time (conditional on no default) plus the payoff to equity in case of default:

$$E_0(V_0) = \mathbb{E}^B\left[\int_0^T e^{-rt}F_t^E I_{\{\tau \not< t\}}dt\right] + \mathbb{E}^B\left[e^{-r\tau}\varphi^E V_\tau I_{\{\tau \leq T\}}\right]. \tag{4.19}$$

In their Proposition 1, Ericsson and Reneby (2001, p. 13) give the value of equity as follows (adapted for our notation):

Proposition 4.2. *The price of equity is given by*

$$\begin{aligned} E_0(V_0) = V &- l_0 \cdot G_{lI}^{\rho}(V,\infty,l_0,\rho) - C \cdot U_{lO}(V,\infty,l_0,\rho) \\ &+ \zeta C \cdot U_{lO}^{\rho}(V,\infty,l_0,\rho) + \varphi^D \frac{C}{r}\left(G_{lI}^{\rho}(V,\infty,l_0,\rho)\right. \\ &\left.- G_{lI}(V,\infty,l_0,\rho)\right) + \varphi^E l_0 G_{lI}^{\rho}(V,\infty,l_0,\rho). \end{aligned} \tag{4.20}$$

Proof. See Ericsson and Reneby (2001, p. 45f.). □

The majority of the 2001 article, however, is on debt valuation. Following the derivation of a valuation formula for a straight bond, the authors discuss previous criticism of firm value based models for not being able to generate yield spreads in line with observable spreads. They argue that this criticism can mainly be attributed to implementation and terminology issues ("model yield" vs. yield calculated from model prices), and that *prices* generated by their model are in line with observed prices (which is what ultimately is important).

The remainder of the paper deals with the problem of estimating the model, and an application to US data. The main conclusions are that bond prices predicted from their model are unbiased and that the model is able to explain a significant portion of yield spread changes.

4.3 Additional Building Blocks within the Probabilistic Framework

For later reference, the values of the following building blocks neither provided by Ericsson and Reneby (1998), nor by Ericsson and Reneby (2001), will be useful:

Definition 4.7. A *down-and-out unit stream* $U_{lO}(V, T, l_0, \rho)$ on V with expiration time T and exponential barrier l_t is defined via the following payoff rate function:

$$\Gamma_t(U_{lO}(V, T, l_0, \rho)) = \Gamma_t(H_{lO}(V, 0, t, l_0, \rho)) \quad t \in [0, T].$$

Lemma 4.12. *The time $t = 0$ value of a down-and-out unit stream as defined above is given by*

$$U_{lO}(V, T, l_0, \rho) = \frac{1}{r}(1 - G_{lI}(V, T, l_0, \rho) - H_{lO}(V, 0, T, l_0, \rho)). \tag{4.21}$$

For infinite maturity, this expression simplifies to

$$U_{lO}(V, \infty, l_0, \rho) = \frac{1}{r}(1 - G_{lI}(V, \infty, l_0, \rho)). \tag{4.22}$$

This proposition basically tells us that the pricing formula for the down-and-out unit stream in the case of an exponential barrier corresponds to the constant barrier formula, when the constant barrier building blocks (unit down-and-in claim and down-and-out heaviside) are replaced by their exponential barrier counterparts (which is what intuition suggests).

For completeness, we give also the proof which can be found (for the constant barrier case) in Ericsson and Reneby (1998, p. 161f.).

Proof.

$$\begin{aligned}
U_{lO}(V, T, l_0, \rho) &= \int_0^T H_{lO}(V, 0, t, l_0, \rho)dt \\
&= \int_0^T e^{-rt}\mathbb{Q}^B\{\tau \not\leq t\}dt \\
&= \left[-\frac{e^{-rt}}{r}\mathbb{Q}^B\{\tau \not\leq t\}\right]_0^T - \int_0^T \frac{e^{-rt}}{r} f^B(X_0, t)dt \\
&= \frac{1}{r}\left(1 - H_{lO}(V, 0, T, l_0, \rho) - G_{lI}(V, T, l_0, \rho)\right),
\end{aligned}$$

where $f^B(X_0, t)$ denotes the first passage time density at t of the process (X_t) to zero (under $\mathbb{Q}^B$). □

Definition 4.8. A *down-and-out asset stream* $O_{lO}(V, T, l_0, \rho)$ on V with expiration time T and exponential barrier l_t is defined via the following payoff rate function:

$$\Gamma_t(O_{lO}(V, T, l_0, \rho)) = \Gamma_t(C_{lO}(V, 0, t, l_0, \rho)) \quad t \in [0, T].$$

Lemma 4.13. *The time* $t = 0$ *value of an asset stream as defined above is given by*

$$O_{lO}(V, T, l_0, \rho) = \frac{1}{\beta}(V_0 - l_0 \cdot G_{lI}^{\rho}(V, T, l_0, \rho) - \Omega_{lO}(V, T, l_0, \rho)). \quad (4.23)$$

For infinite maturity, this expression simplifies to

$$O_{lO}(V, \infty, l_0, \rho) = \frac{1}{\beta}(V - l_0 \cdot G_{lI}^{\rho}(V, \infty, l_0, \rho)). \quad (4.24)$$

Compared to the down-and-out unit stream, the formula for the down-and-out asset stream in the exponential barrier case differs more from the constant barrier case formula, as $G_{LI}(\cdot)$ is not replaced by $G_{lI}(\cdot)$, but by $G_{lI}^{\rho}(\cdot)$ (the "barrier in- or decreasing unit down-and-in claim"). We adapt the constant-barrier case proof given in Ericsson and Reneby (1998, p. 162 f.) to show where this effect comes in.

Proof.

$$\begin{aligned}
O_{lO}(V, T, l_0, \rho) &= \int_0^T C_{lO}(V, 0, t, l_0, \rho)dt && (4.25)\\
&= \mathbb{E}^B\left[\left[\int_0^T e^{-rt}V_t I_{\{\tau \not\leq t\}}\right] dt\right. && (4.26)\\
&= \int_0^T e^{-rt}\mathbb{E}^B[V_t]\mathbb{E}^V[I_{\{\tau \not\leq t\}}]dt && (4.27)\\
&= \int_0^T e^{-\beta t}V_0\mathbb{Q}^V\{\tau \not\leq t\}dt && (4.28)\\
&= V_0\left[-\frac{e^{-\beta t}}{\beta}(1 - \mathbb{Q}^V\{\tau \not\leq t\})\right]_0^T \\
&\quad - V_0\int_0^T \frac{e^{-\beta t}}{\beta} f^V(X_0, t)dt, && (4.29)
\end{aligned}$$

where $f^V(X_0, t)$ denotes the first passage time density at t of the process (X_t) to zero (under $\mathbb{Q}^V$). The first term in equation (4.29) can be simplified and rewritten as

$$V_0\left[-\frac{e^{-\beta t}}{\beta}(1-\mathbb{Q}^V\{\tau \not\leq t\})\right]_0^T = \frac{1}{\beta}(V_0 - \Omega_{lO}(V, T, l_0, \rho)).$$

For the second term, we note (following Ericsson and Reneby (1998, p. 163) and adapting to the exponential barrier case) that

$$\begin{aligned} V_0\int_0^T e^{-\beta t} f^V(X_0, t)dt &= l_t\int_0^T e^{-rt} f^B(X_0, t)dt \\ &= l_0\int_0^T e^{(\rho - r)t} f^B(X_0, t)dt \\ &= l_0 G^{\rho}_{lI}(V, \infty, l_0, \rho)). \end{aligned}$$

□

This shows that care should be taken when extending results from the literature that have been derived for a constant barrier to the case of an exponential barrier, especially whenever parts of claims depend on the value of the barrier at a certain time.

For the modelling of sinking fund-like payments, the following securities will prove useful:

Definition 4.9. An exponentially in- or decreasing unit stream $U^{\nu}(T)$ with maturity T is a security with the following payoff rate function:

$$\Gamma_t(U^{\nu}(T)) = e^{\nu t} \quad t \in [0, T]. \tag{4.30}$$

Graphically, the payoff rate of this claim is similar to that of the barrier-increasing unit stream shown in Figure 4.5 (with the additional flexibility that ν, the growth rate of the payment stream, need not be equal to ρ, the growth rate of the barrier, and – for the no-default case – without the down-and-out feature).

For completeness, we start with the pricing formula for the exponentially in- or decreasing unit stream in the no-default case:

Lemma 4.14. *The price of the exponentially in- or decreasing unit stream as defined above is (for $\nu \neq r$) given by*

$$U^{\nu}(T) = \frac{1 - e^{-(r-\nu)T}}{r - \nu}. \tag{4.31}$$

For $\nu = r$, $U^{\nu}(T) = T$. *For infinite maturity and* $\nu < r$, *equation* (4.31) *converges to*

$$U^{\nu}(\infty) = \frac{1}{r - \nu}.$$

For $\nu \geq r$ *and* $T \to \infty$, *the price does not exist.*

Proof. Follows from equation (4.30) by straightforward integration (see proof of Lemma 4.18). □

The derivation of a pricing formula for the down-and-out counterpart of the exponentially in- or decreasing unit stream is simplified by the introduction of another security:

Definition 4.10. An *exponentially in- or decreasing unit down-and-in claim* $G^{\nu}_{UI}(V, T, l_0, \rho)$ on V with expiration time T, growth rate ν and exponential barrier l_t is a security with the following payoff function:

$$\Gamma_{\tau}(G^{\nu}_{UI}(V, T, l_0, \rho)) = e^{\nu\tau} I_{\{\tau \leq T\}}.$$

Graphically, the payoff of this claim is similar to that of the barrier-increasing unit down-and-in claim shown in Figure 4.4 (with the additional flexibility that ν, the growth rate of the payment in case of default, need not be equal to ρ, the growth rate of the barrier).

An inspection of the derivation of the "barrier-increasing unit down-and-in-claim" (cf. Lemma 4.10 in Reneby (1998, p. 79)) shows that this claim can easily be generalized to "interest rates" ν which are different from the growth rate of the barrier. To show this, we make use of a Lemma provided by Ericsson and Reneby (1998, p. 160):

Lemma 4.15 (Ericsson and Reneby (1998)).

$$\int_0^T e^{-\kappa t} f^n(X_0, t) dt = e^{(\mu^m_X - \mu^n_X) X_0} (1 - \mathbb{Q}^m\{\tau \nleq T\}),$$

where $f^n(\cdot)$ *denotes the first passage time density at* t *of the process* (X_t) *to zero (under measure* $\mathbb{Q}^n$*) and*

$$\mu^m_X = \sqrt{(\mu^n_X)^2 + 2\kappa}.$$

This result holds if the discriminant is positive.

Proof. See Ericsson and Reneby (1998, p. 161). □

Armed with this lemma, we can now extend Lemma 4.10 by generalizing this proof:

Lemma 4.16. *The time* $t = 0$ *price of an exponentially in- or decreasing unit down-and-in claim as defined above is given by*

$$G^{\nu}_{lI}(V,T,l_0,\rho) = G^{\nu}_{lI}(V,\infty,l_0,\rho)\left(1 - \mathbb{Q}^{G^{\nu}}\{\tau \not\leq T, V_T > l_T\}\right), \tag{4.32}$$

where $G^{\nu}_{lI}(V,\infty,l_0,\rho)$ *is given by*

$$G^{\nu}_{lI}(V,\infty,l_0,\rho) = \left(\frac{V}{l_0}\right)^{-\theta(r-\nu)},$$

and $\mathbb{Q}^{G^{\nu}}$ *is the martingale measure with* $G^{\nu}_{lI}(V,\infty,l_0,\rho)$ *as numeraire. The associated Girsanov kernel (i.e., the drift of* (X_t) *under* $\mathbb{Q}^{G^{\nu}}$*) is* $\mu^{G^{\nu}}_X = \mu^B_X - \theta(r-\nu)\sigma$*. The result holds if* $(\mu^B_X)^2 \geq -2(r-\nu)$*.*

Proof. Standard martingale valuation theory tells us that

$$G^{\nu}_{lI}(V,T,l_0,\rho) = \int_0^T e^{-(r-\nu)} f^B(X_0,t)dt.$$

The result in equation (4.32) follows immediately from Lemma 4.15. □

Remark 4.2. The sign of ν determines whether the claim is in- or decreasing. Note that, for $\nu = \rho$, Lemma 4.16 provides the extension of the claim in Lemma 4.10 to the finite maturity case.

Remark 4.3. Note that the definition of this claim is *relative to time* 0. Thus, both the potential payoff and the value of the claim are time-dependent.

Remark 4.4. Note that if $(\mu^B_X)^2 \geq -2(r-\nu)$ is not satisfied, the price of the *finite* version of the claim still exists. However, it cannot be calculated using equation (4.32), since this equation essentially relies on calculating the value of the finite claim as the difference between two infinite claims, one starting at time $t = 0$, the other starting at time $t = T$. This is definitely a limitation of this approach, although we never encountered any problems for realistic parameter values in the numerical applications in Chapters 9 and 10.

Definition 4.11. An *exponentially in- or decreasing down-and-out unit stream* $U^{\nu}_{lO}(V, T, l_0, \rho)$ on V with expiration time T, growth rate ν and exponential barrier l_t is a security with the following payoff rate function:

$$\Gamma_t(U^{\nu}_{lO}(V, T, l_0, \rho)) = e^{\nu t}\Gamma_t(H_{lO}(V, 0, t, l_0, \rho)) \quad t \in [0, T].$$

Lemma 4.17. *The time* $t = 0$ *price of an exponentially in- or decreasing down-and-out unit stream as defined above is given by*

$$U^{\nu}_{lO}(V, T, l_0, \rho) = \frac{1}{r - \nu}\left(1 - e^{\nu T} H_{lO}(V, 0, T, l_0, \rho) \right. \\ \left. - G^{\nu}_{lI}(V, T, l_0, \rho)\right). \tag{4.33}$$

For infinite maturity, this converges to

$$U^{\nu}_{lO}(V, \infty, l_0, \rho) = \frac{1}{r - \nu}\left(1 - G^{\nu}_{lI}(V, \infty, l_0, \rho)\right). \tag{4.34}$$

The result holds if $r > \nu$. *If* $r = \nu$, *the price is found as the limiting case.*

Proof. From the definition of this claim and standard martingale valuation theory, we have

$$\begin{aligned} U^{\nu}_{lO}(\cdot) &= \mathbb{E}^B\left[\int_0^T e^{-(r-\nu)t} H_{lO}(V, 0, t, l_0, \rho) dt\right] \\ &= \left[-\frac{e^{-(r-\nu)t}}{r-\nu}\mathbb{Q}^B\{\tau \not\leq t\}\right]_0^T - \int_0^T \frac{e^{-(r-\nu)t}}{r-\nu} f^B(X_0, t) dt. \end{aligned}$$

Equation (4.33) follows by remembering that

$$\int_0^T e^{-(r-\nu)t} f^B(X_0, t) dt = G^{\nu}_{lI}(V, T, l_0, \rho)$$

and

$$\begin{aligned} e^{-(r-\nu)T}\mathbb{Q}^B\{\tau \not\leq T\} &= e^{-(r-\nu)T}\mathbb{Q}^B\{\tau \not\leq T, V_T > l_T\} \\ &= e^{\nu T} H_{lO}(V, 0, T, l_0, \rho). \end{aligned}$$

Equation (4.34) follows as the limiting case. □

Remark 4.5. Note that this proposition, for $\nu = \rho$, contains the barrier-increasing unit stream from Lemma 4.11 as a special case.

Remark 4.6. If $r < \nu$, the existence of a price depends on the expressions in brackets (e.g., in the infinite case, we need that $G_{\mathrm{U}}^{\gamma}(\cdot) > 1$). This condition depends on the input variables in a non-trivial way. A necessary (but not sufficient) condition for the existence of a price in the infinite maturity case is $(\mu_X^B)^2 \geq -2(r - \nu)$.

In the finite maturity case, a price always exists, but may not be computable using equation (4.33) if the price of one of the components does not exist (cf. Remark 4.4).

For sinking-fund-like payment structures which are not exponential, but linear, a linearly decreasing unit stream will be useful.

Definition 4.12. A linearly decreasing unit stream ${}^{\mathrm{dec}}_{\mathrm{lin}}\mathrm{U}(T)$ with expiration time T is a security with the following payoff rate function:

$$\Gamma_t({}^{\mathrm{dec}}_{\mathrm{lin}}\mathrm{U}(T)) = \frac{T - t}{T} \qquad t \in [0, T].$$

The payoff rate of this security over time is illustrated in Figure 4.6.

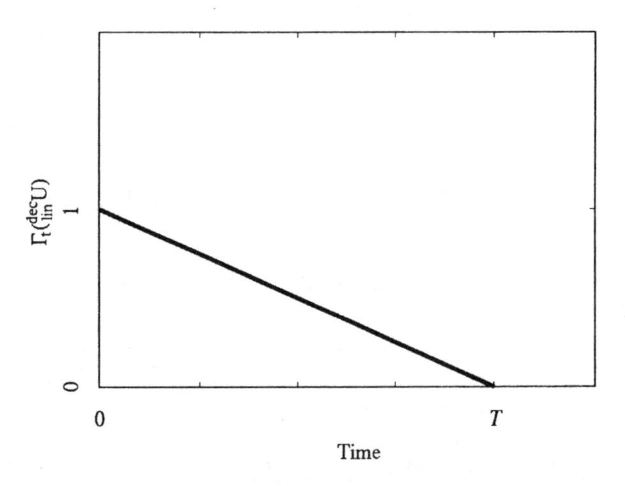

Figure 4.6: Payoff rate function of a linearly decreasing unit stream

For completeness, we start with the pricing formula of a linearly decreasing unit stream in the no-default case ($L = 0$):

Lemma 4.18. *The time* $t = 0$ *price of the linearly decreasing unit stream as defined above is given by*

$$ {}^{\text{dec}}_{\text{lin}}\mathcal{U}(T) = \frac{rT + e^{-rT} - 1}{r^2 T}. $$

Proof.

$$ \begin{aligned} {}^{\text{dec}}_{\text{lin}}\mathcal{U}(V,T) &= \int_0^T e^{-rt}\Big(1 - \frac{t}{T}\Big)dt \\ &= \int_0^T e^{-rt}dt - \frac{1}{T}\int_0^T e^{-rt}tdt \\ &= \int_0^T e^{-rt}dt - \frac{1}{rT}\int_0^T e^{-rt}dt + \frac{1}{rT}e^{-rT}T \\ &= \frac{1 - e^{-rT}}{r} - \frac{1 - e^{-rT}}{r^2T} + \frac{e^{-rT}}{r} \\ &= \frac{rT + e^{-rT} - 1}{r^2T}. \end{aligned} $$

□

The derivation of a pricing formula for the down-and-out counterpart of the linearly decreasing unit stream is simplified by the introduction of another security:

Definition 4.13. A *linearly decreasing unit down-and-in claim* ${}^{\text{dec}}_{\text{lin}}\mathcal{G}_{\mathcal{U}}(V, T, l_0, \rho)$ on V with expiration time T and exponential barrier l_t is a security with the following payoff function:

$$ \Gamma_\tau({}^{\text{dec}}_{\text{lin}}\mathcal{G}_{\mathcal{U}}(V, T, l_0, \rho)) = \frac{T - \tau}{T}\mathbb{I}_{\{\tau \le T\}} \left(= \Gamma\left(\frac{T-\tau}{T}\mathcal{G}_{\mathcal{U}}(V, T, l_0, \rho)\right)\right). $$

The payoff of a linearly decreasing unit down-and-in claim depends on the time bankruptcy actually occurs: The earlier bankruptcy occurs (the smaller τ), the larger the payoff. If bankruptcy does not occur in the time interval $[0, T]$, the claim expires worthless. Possible payoffs of a linearly decreasing unit down-and-in claim for various bankruptcy times are illustrated in Figure 4.7.

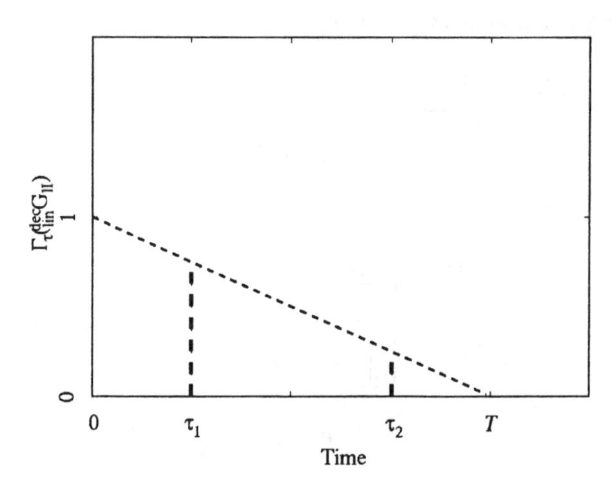

Figure 4.7: Possible payoffs of a linearly decreasing unit down-and-in claim for various bankruptcy times τ_i

Lemma 4.19. *The time* $t = 0$ *price of the linearly decreasing unit down-and-in claim as defined above is given by*

$$ {}^{\text{dec}}_{\text{lin}}G_{\text{II}}(V, T, l_0, \rho) = \frac{1}{\mu_X^G\sqrt{T}}\left(\left(\frac{V}{l_0}\right)^{\frac{-\mu_X^G-\mu_X^B}{\sigma}} N(q_1(T))q_1(T)\right.$$

$$\left. - \left(\frac{V}{l_0}\right)^{\frac{\mu_X^G-\mu_X^B}{\sigma}} N(q_2(T))q_2(T)\right)$$

with μ_X^B *and* μ_X^G *as defined in equation* (4.16), *and*

$$q_1(t) = \frac{\ln\frac{l_0}{V}}{\sigma\sqrt{t}} + \mu_X^G\sqrt{t}$$

and

$$q_2(t) = \frac{\ln\frac{l_0}{V}}{\sigma\sqrt{t}} - \mu_X^G\sqrt{t}.$$

Proof. See Leland and Toft (1996, Appendix A). Note that $J(T)$ in their notation corresponds exactly to our linearly decreasing unit down-and-in claim. They derive the value of this claim under the assumption of a constant barrier. The extension to the exponential barrier case can easily be achieved by adjusting the drifts μ_X^m accordingly. □

Definition 4.14. A *linearly decreasing down-and-out unit stream* ${}^{dec}_{lin}U_{lO}(V, T, l_0, \rho)$ on V with expiration time T and exponential barrier l_t is a security with the following payoff rate function:

$$\Gamma_t({}^{dec}_{lin}U_{lO}(V, T, l_0, \rho)) = \frac{T-t}{T}\Gamma_t(H_{lO}(V, 0, t, l_0, \rho)) \quad t \in [0, T].$$

The payoff rate of this security over time is illustrated in Figure 4.8: Before bankruptcy occurs, it pays off like a linearly decreasing unit stream, while as soon as bankruptcy occurs, the payoff rate drops to zero and stays there.

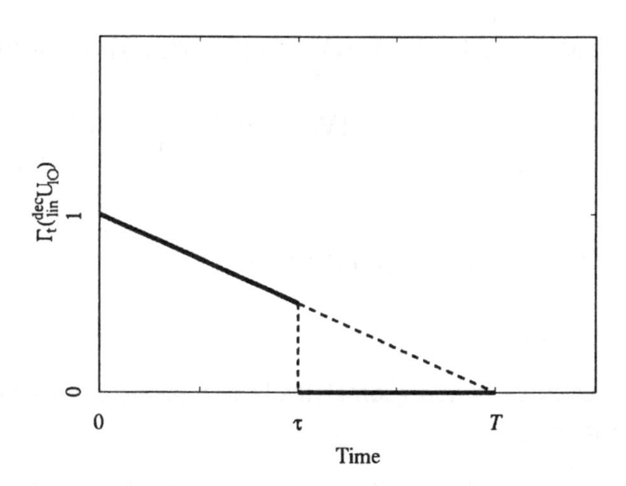

Figure 4.8: Payoff rate of a linearly decreasing down-and-out unit stream over time

Lemma 4.20. *The time* $t = 0$ *price of the linearly decreasing down-and-out unit stream as defined above is given by*

$$ {}^{dec}_{lin}U_{lO}(V, T, l_0, \rho) = \frac{1}{r}\left(1 - \frac{1}{T}U_{lO}(V, T, l_0, \rho) - {}^{dec}_{lin}G_{lI}(V, T, l_0, \rho)\right). \quad (4.35)$$

Proof. From the definition of this claim, we have

$$
\begin{aligned}
{}^{\text{dec}}_{\text{lin}}U_{l0}(V,T,l_0,\rho) &= \int_0^T \left(\left(1-\frac{t}{T}\right) H_{l0}(V,0,t,l_0,\rho)\right) dt & (4.36)\\
&= \int_0^T H_{l0}(V,0,t,l_0,\rho)dt - \frac{1}{T}\int_0^T tH_{l0}(V,0,t,l_0,\rho)dt & (4.37)\\
&= U_{l0}(V,T,l_0,\rho) \\
&\quad - \frac{1}{T}\left([tU_{l0}(V,t,l_0,\rho)]_0^T - \int_0^T U_{l0}(V,t,l_0,\rho)dt\right) & (4.38)\\
&= \frac{1}{T}\int_0^T U_{l0}(V,t,l_0,\rho)dt, & (4.39)
\end{aligned}
$$

where equation (4.38) comes from integration by parts. From equation (4.21), we have

$$
\begin{aligned}
\frac{1}{T}\int_0^T U_{l0}(\cdot)dt &= \frac{1}{rT}\int_0^T (1 - G_{ll}(V,t,l_0,\rho) - H_{l0}(V,0,t,l_0,\rho))\,dt \\
&= \frac{1}{r} - \frac{1}{rT}\int_0^T G_{ll}(V,t,l_0,\rho)dt - \frac{1}{rT}U_{l0}(V,T,l_0,\rho).
\end{aligned}
$$

Noting that

$$
\frac{1}{T}\int_0^T G_{ll}(V,t,l_0,\rho)dt = {}^{\text{dec}}_{\text{lin}}G_{ll}(V,T,l_0,\rho),
$$

equation (4.35) follows. □

In addition to linearly and geometrically decreasing claims, we will also need a hybrid form:

Definition 4.15. A *linearly decreasing, exponentially in- or decreasing unit stream* ${}^{\text{dec}}_{\text{lin}}U^{\nu}(T)$ with maturity T is a security with the following payoff rate function:

$$
\Gamma({}^{\text{dec}}_{\text{lin}}U^{\nu}(T)) = \frac{T-t}{T}e^{\nu t} \quad t \in [0,T].
$$

The payoff rate of this security over time is illustrated in Figure 4.9.

For completeness, we start again with the pricing formula of a linearly decreasing, exponentially in- or decreasing unit stream in the no-default case ($L = 0$):

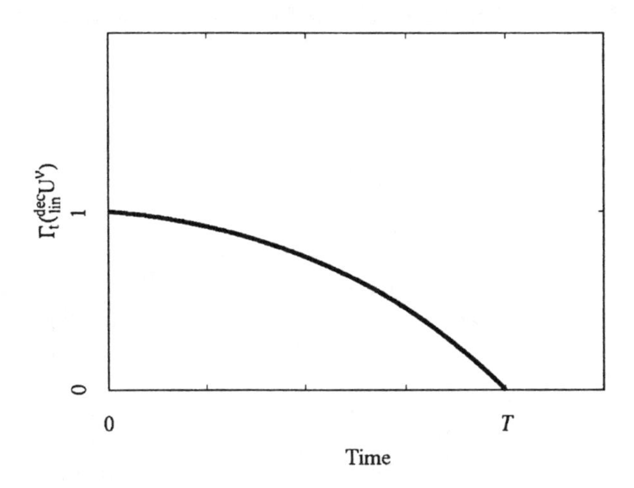

Figure 4.9: Payoff rate function of a linearly decreasing, exponentially in- or decreasing unit stream

Lemma 4.21. *The time* $t = 0$ *price of the linearly decreasing, exponentially in- or decreasing unit stream as defined above is (for* $\nu \neq r$*) given by*

$$ {}^{\text{dec}}_{\text{lin}}U^{\nu}(T) = \frac{(r-\nu)T + e^{-(r-\nu)T} - 1}{r^2 T}. $$

For $\nu \to r$*, the price converges to* $T/2$.

Proof. Along the lines of the proof of Lemma 4.18. □

The derivation of a pricing formula for the down-and-out counterpart of the linearly decreasing, exponentially in- or decreasing unit stream is simplified by the introduction of another security:

Definition 4.16. A *linearly decreasing, exponentially in- or decreasing unit down-and-in claim* ${}^{\text{dec}}_{\text{lin}}G^{\nu}_{\text{ll}}(V, T, l_0, \rho)$ on V with expiration time T, growth rate ν, and exponential barrier l_t, is a security with the following payoff function:

$$ \Gamma_\tau({}^{\text{dec}}_{\text{lin}}G^{\nu}_{\text{ll}}(V, T, l_0, \rho)) = \frac{T-\tau}{T} e^{\nu\tau} I_{\{\tau \leq T\}} \left(= \Gamma_\tau \left(\frac{T-\tau}{T} G^{\nu}_{\text{ll}}(V, T, l_0, \rho) \right) \right). $$

Possible payoffs of a linearly decreasing, exponentially in- or decreasing unit down-and-in claim for various bankruptcy times are illustrated in Figure 4.10.

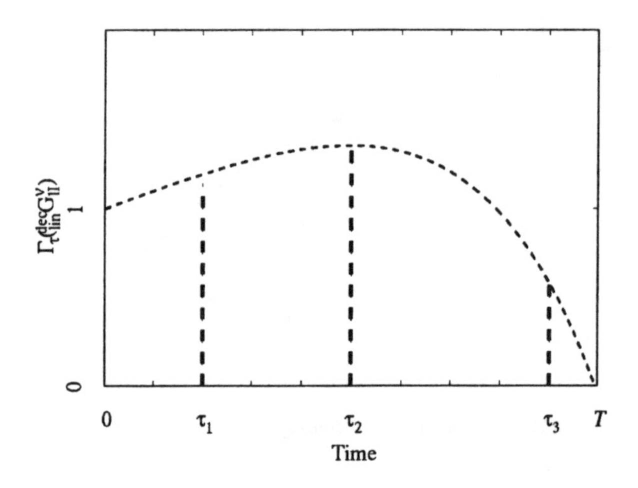

Figure 4.10: Possible payoffs of a linearly decreasing, exponentially in- or decreasing unit down-and-in claim for various bankruptcy times τ_i

Lemma 4.22. *The time* $t = 0$ *price of the linearly decreasing, exponentially in- or decreasing unit down-and-in claim as defined above is given by*

$$
{}^{\text{dec}}_{\text{lin}}G^{\nu}_{II}(V, T, l_0, \rho) = \frac{1}{\mu_X^{G^\nu}\sqrt{T}} \left(\left(\frac{V}{l_0}\right)^{\frac{-\mu_X^{G^\nu} - \mu_X^B}{\sigma}} N(q_1(T))q_1(T) \right.
$$

$$
\left. - \left(\frac{V}{l_0}\right)^{\frac{\mu_X^{G^\nu} - \mu_X^B}{\sigma}} N(q_2(T))q_2(T) \right)
$$

with μ_X^B *and* $\mu_X^{G^\nu}$ *as defined before, and*

$$
q_1(t) = \frac{\ln \frac{l_0}{V}}{\sigma\sqrt{t}} + \mu_X^{G^\nu}\sqrt{t}
$$

and

$$q_2(t) = \frac{\ln \frac{l_0}{V}}{\sigma\sqrt{t}} - \mu_X^{G^\nu}\sqrt{t}.$$

The result holds if $(\mu_X^B)^2 \geq -2(r-\nu)$.

Proof. This claim is a straightforward extension of the linearly decreasing unit down-and-in claim in Lemma 4.19. The only necessary change is from μ_X^G to $\mu_X^{G^\nu}$ throughout. □

Definition 4.17. A *linearly decreasing, exponentially in- or decreasing down-and-out unit stream* ${}^{dec}_{lin}U^{\nu}_{lO}(V, T, l_0, \rho)$ on V with expiration time T, growth rate ν, and exponential barrier l_t, is a security with the following payoff rate function:

$$\Gamma_t({}^{dec}_{lin}U^{\nu}_{lO}(V, T, l_0, \rho)) = \frac{T-t}{T} e^{\nu t}\Gamma_t(H_{lO}(V, 0, T, l_0, \rho)) \quad t \in [0, T].$$

The payoff rate of this security over time is illustrated in Figure 4.11: Before bankruptcy occurs, it pays off like a linearly decreasing unit stream, while as soon as bankruptcy occurs, the payoff rate drops to zero and stays there.

Lemma 4.23. *The time* $t = 0$ *price of the linearly decreasing, exponentially in- or decreasing down-and-out unit stream as defined above is given by*

$$ {}^{dec}_{lin}U^{\nu}_{lO}(V, T, l_0, \rho) = \frac{1}{(r-\nu)}\left(1 - \frac{1}{T}U^{\nu}_{lO}(V, T, l_0, \rho) - {}^{dec}_{lin}G^{\nu}_{lI}(V, T, l_0, \rho)\right). \tag{4.40}$$

The result holds if $r > \nu$. *If* $r = \nu$, *the price is found as the limiting case.*

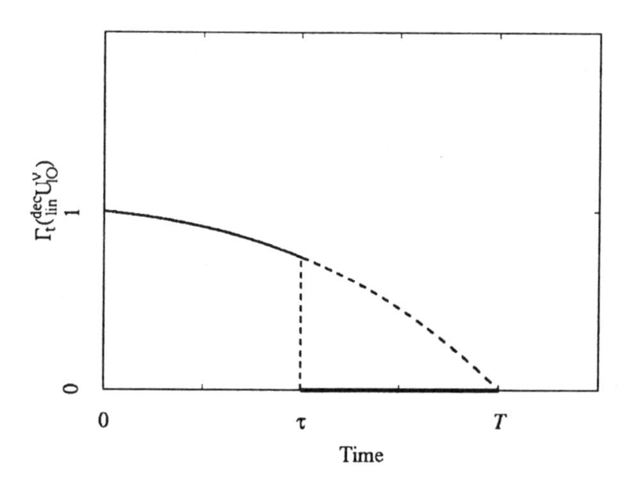

Figure 4.11: Payoff rate of a linearly decreasing, exponentially increasing down-and-out unit stream over time

Proof. From the definition, we have

$$
\begin{aligned}
{}^{dec}_{lin}U^{\nu}_{lO}(\cdot) &= \int_0^T \left(\left(1 - \frac{t}{T}\right) e^{\nu t} H_{lO}(V, 0, t, l_0, \rho) \right) dt && (4.41)\\
&= \int_0^T e^{\nu t} H_{lO}(V, 0, t, l_0, \rho) dt - \frac{1}{T} \int_0^T t e^{\nu t} H_{lO}(V, 0, t, l_0, \rho) dt && (4.42)\\
&= U^{\nu}_{lO}(V, T, l_0, \rho) \\
&\quad - \frac{1}{T} \left([t U^{\nu}_{lO}(V, t, l_0, \rho)]_0^T - \int_0^T U^{\nu}_{lO}(V, t, l_0, \rho) dt \right) && (4.43)\\
&= \frac{1}{T} \int_0^T U^{\nu}_{lO}(V, t, l_0, \rho) dt, && (4.44)
\end{aligned}
$$

where equation (4.43) comes from integration by parts. From equation

(4.33), we have

$$\frac{1}{T}\int_0^T U_{lO}^{\gamma}(\cdot)dt = \frac{\int_0^T \left(1 - G_{lI}^{\gamma}(V,t,l_0,\rho) - e^{\nu T}H_{lO}(V,0,t,l_0,\rho)\right) dt}{(r-\nu)T}$$
$$= \frac{\left(1 - \frac{1}{T}\int_0^T G_{lI}^{\gamma}(V,t,l_0,\rho)dt - \frac{1}{T}U_{lO}^{\gamma}(V,T,l_0,\rho)\right)}{(r-\nu)}.$$

Noting that

$$\frac{1}{T}\int_0^T G_{lI}^{\gamma}(V,t,l_0,\rho)dt = {}_{\text{lin}}^{\text{dec}}G_{lI}^{\gamma}(V,T,l_0,\rho),$$

equation (4.40) follows. □

Remark 4.7. For $r < \nu$, compare Remark 4.6.

Chapter 5

A Review of Firm Value Based Security Pricing Models from a Probabilistic Perspective

In this chapter, we will review some well-known models from the literature on firm value based pricing of corporate securities. We do not strive to cover each and every pricing model that can be found in the literature. Instead, we focus on "true" firm value based models, where the stock price process explicitly and directly depends on the firm's capital structure. Therefore, so-called intensity based credit risk models like Jarrow and Turnbull (1995), Jarrow, Lando, and Turnbull (1997) or Duffie and Singleton (1997, 1999) are excluded. From the remaining models (also known as "structural credit risk models"), we select according to the criteria of *importance* (in the sense of being influential for further work) and *existence of closed-form solutions.* Numerical results for most of the models discussed will be provided in Chapters 9 and 10.

We present these classical models within the Ericsson and Reneby (1998, 2001) framework to show that

- most of the models reviewed are nested by this framework (in fact, all of the models that assume a constant risk-free interest rate except some features of the Leland (1998) model),

- all those models can be derived conveniently within this framework

- and economic interpretation is greatly simplified compared to many formulae in the literature that have been derived using the PDE approach.

5.1 Finite-Maturity Discount Bonds, No Intermediate Default (Merton 1974)

Whereas the general guidelines have already been laid down by Black and Scholes (1973), the paper by Merton (1974) was the first to provide a formal exposition of the application of contingent claims valuation techniques to the pricing of corporate securities. For a description of the model and some important results, compare Section 2.3. We note that the Merton (1974) model can be viewed as a special case of the Ericsson and Reneby (1998) model with no taxes, no deviations from absolute priority, no intermediate bankruptcy, and a very primitive capital structure consisting only of equity and discount bonds, respectively.

First, Merton studies the value of a firm with this simple capital structure. Following the derivation of the respective pricing formulae given in equations (2.20) and (2.21), he studies the (default) risk structure of corporate bond yields. Then, he notes that the well-known Modigliani and Miller (1958) theorem also holds in the presence of bankruptcy in his model. The proof for this claim depends on the possibility of replicating a risky bond by trading in equity of an otherwise similar company and a riskless bond (note the relation to the discussion of Assumption 4.1 on page 60 above). Holding everything else constant, Merton moves on to examine required equity and bond returns for varying leverage levels.

Finally, Merton shows how to price perpetual coupon bonds (both callable and non-callable) in his extended Black–Scholes model. Whereas he provides an analytic solution for the non-callable perpetual coupon bond (albeit using different assumptions regarding financing of coupon payments by asset sales), he recommends resorting to numerical methods to handle the case where the bond is callable by the firm.

5.2 Finite-Maturity, Continuous-Coupon Bonds, Intermediate Default (Black and Cox 1976)

Black and Cox (1976) extend Merton's work by investigating the effects of various bond indenture provisions. They start by noting that Merton's model implies several specific assumptions regarding the bond indenture that restrict the firm's investment, payout and further financing policies. In particular, they extend Merton's model by allowing for financial distress to occur not only at maturity of the debt, but also during the bond's lifetime, criticizing that

> "...it [Merton's model] assumes that the fortunes of the firm may cause its value to rise to an arbitrarily high level or dwindle to nearly nothing without any sort of reorganization occurring in the firm's financial arrangements." (p. 352)

Black and Cox (1976) start by considering so-called *safety covenants.* These are provisions which give the bondholders the right to force the firm into some form of reorganization as soon as the firm value reaches a pre-specified lower level. This level is modelled as a deterministic, exponential function of time. Thus, the Black and Cox (1976) model can be written as follows:

$$dV_t = (\mu_t - \beta)V_t dt + \sigma_V V_t dW_t, \tag{5.1}$$

$$E_T = (V_T - D)_+, \tag{5.2}$$

$$D_T = \min(D, V_T), \tag{5.3}$$

$$E_\tau = 0, \tag{5.4}$$

and

$$D_\tau = l_0 e^{\rho\tau}, \tag{5.5}$$

where $l(\rho, t) = l_0 e^{\rho t}$ represents the exponential barrier at time t, and τ is the first passage time of the firm value to this barrier. Black and Cox (1976) implicitly assume that $l_0 e^{\rho t} \leq De^{-r(T-t)} \quad \forall t$ (this follows from $E_\tau = 0$ if bondholders should not receive more than they are owed in case of default).

The Black and Cox (1976) model is a special case of the Ericsson and Reneby (1998) model with zero tax rate, and no deviations from absolute priority. Contrary to the Merton (1974) model, however, it features intermediate bankruptcy, triggered by an exponential barrier, and a constant asset payout ratio β.

Therefore, equity can here be viewed as a down-and-out call on the firm's value V plus β down-and-out asset streams, both with an exponential barrier (the third component, coming from the payoff in case of default, is zero due to the restriction on the barrier given above):

$$E_0 = C_{l0}(V_0, D, T, l_0, \rho) + \beta O_{l0}(V_0, T, l_0, \rho).$$

Using $V_0 = E_0 + D_0$, we get for the bond:

$$D_0 = V_0 - C_{l0}(V_0, D, T, l_0, \rho) - \beta O_{l0}(V_0, T, l_0, \rho).$$

Compared to the rather messy looking formula given derived by Black and Cox (1976, p. 356), our formula is much more amenable to economic interpretation and analysis. Both equations here can be expanded using the corresponding formulae for the building blocks defined above.

Black and Cox (1976, p. 364) also note that, in model variants where equity makes intermediate payments and in the absence of a barrier imposed by bond covenants, there exists a level of the barrier L (call it L_{end}) that maximizes the value of equity. If (V_t) fell below this "endogenous bankruptcy level", equityholders would be unwilling to contribute further payments and would surrender the firm to the bondholders. For the case of a perpetual bond with continuous coupon payments at a rate of C per year, they derive the value of this endogenous barrier as

$$L_{end} = \frac{C}{r(1 + \frac{\sigma^2}{2r})}. \tag{5.6}$$

Given the difference in the payment structure of a perpetual bond with continuous coupons compared to a finite-maturity discount bond, it should not come as a surprise that this barrier is now *constant* rather than an exponential function of time.

In the remaining part of the paper, Black and Cox (1976) analyze some properties of discount bonds within their setting, followed by a treatment of subordinated bonds and an investigation of effects due to restrictions on the financing of interest and dividend payments.

5.3 Finite-Maturity, Discrete-Coupon Bonds, Intermediate Default (Geske 1977)

Geske (1977) starts with a brief review of the Merton (1974) model. He focuses on the problem of valuing corporate coupon bonds of finite maturity with discrete coupon payments and finds that (up to the time his article was published), "no formula for valuing these risky coupon bonds has been developped,..." (p. 541).

Geske builds upon the insight of Black and Scholes (1973) that coupon bonds can be valued by viewing the common stock as a compound option:

If the coupons are assumed to be financed by stockholders, they have the choice at each coupon date either to pay the coupon, or to default on the coupon and let the bondholders take over the firm. In contrast to Black and Scholes (1973), Geske (1977) also provides an analytical framework that makes the numerical computation of prices of coupon bonds possible. However, his solution is not "genuinely closed-form", but in terms of multivariate normal integrals, where the dimensionality depends on the number of remaining coupons. Therefore, this model is only mentioned here for completeness without a detailed description.

In the remaining part of the paper, Geske (1977) proceeds along the lines of Black and Cox (1976) to investigate the effects of various common bond indenture conditions, and he provides a formula for the valuation of subordinated debt.

5.4 Finite-Maturity, Convertible Discount Bonds, No Intermediate Default (Ingersoll 1977a)

Ingersoll (1977a) examines the pricing of convertible bonds and preferred stocks within the extended Black–Scholes model. He starts by finding general restrictions on rational convertibles pricing. In particular, he derives optimal call policies for the issuing company as well as optimal conversion policies for convertibles holders under rather general assumptions.

5.4.1 Convertible Discount Bonds

The first extension studied in detail is the conversion feature. Whereas Ingersoll (1977a) mainly works with partial differential equations, we will re-derive some of his results using the martingale pricing approach here. The general framework is that of Merton (1974) (see Sections 2.3 and 5.1 for details). In addition, the discount bond can be converted for a fraction γ of the post-conversion equity. γ is often referred to as the "dilution factor" and is given by $\gamma = m/(m+n)$, where n denotes the number of shares of common stock outstanding and m denotes the number of shares issued upon conversion of the bond.

Having shown that conversion is never optimal prior to maturity (Theorem I in Ingersoll (1977a, p. 295)), it is easy to see from Figure 5.1 that the convertible discount bond is, in essence, equivalent to a non-convertible bond (with the same terms) plus γ European call options on V with strike D/γ. Denoting the time t value of the convertible discount bond with face value D by D_t^c, we get (using equation (2.21))

$$\begin{aligned} D_0^c &= D_0 + \gamma C_0(V, D/\gamma, T) \\ &= V_0 - C_0(V, D, T) + \gamma C_0(V, D/\gamma, T). \end{aligned} \tag{5.7}$$

From equation (5.7), we see that the convertible is equal to a portfolio

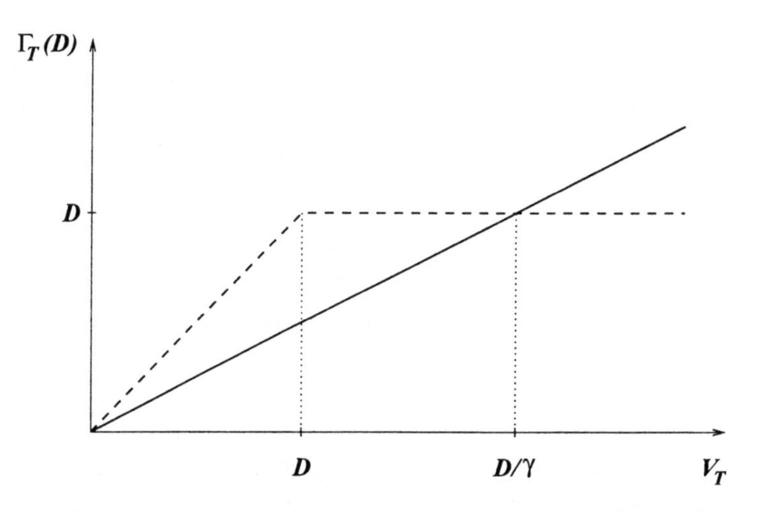

Figure 5.1: Convertible discount bond in the Ingersoll (1977a) model: Payoff to bondholders depending on V_T and the conversion decision. The solid line represents the value in case of conversion, the dotted line the value if not converted.

consisting of one long asset claim, γ long calls with strike D/γ and one short call on firm value with strike D.

Equity can then be valued as

$$\begin{aligned} E_0 &= V_0 - D_0^c \\ &= C_0(V, D, T) - \gamma C_0(V, D/\gamma, T). \end{aligned} \tag{5.8}$$

5.4.2 Callable Convertible Discount Bonds

Ingersoll (1977a) continues with an analysis of convertible discount bonds which are callable, i.e., the firm may decide at any point in time to "call" the bond. The bondholders then have to decide immediately whether they choose to receive the call price or to convert. The call price $k(\cdot)$ is modelled as an exponential function of maturity, similar to the safety covenant in the Black and Cox (1976) model (cf. equation (5.5)): $k(t,\cdot) = De^{-\rho(T-t)}$. This corresponds to a discount bond with face value D and continuously compounded yield ρ.

We will re-derive Ingersoll's valuation formula for a firm with a capital structure consisting only of common stock and convertible debt following a different path. This leads to a more elegant and direct derivation. Theorem IV in Ingersoll (1977a) gives the optimal call strategy for this callable convertible bond: The firm should call the bond at time τ, where τ denotes the first passage time of the firm value V_t to $k(t,\cdot)/\gamma$.

Given this optimal policy, the problem of valuing a callable convertible can be recast in terms of barrier options: As soon as the firm value V_t hits the barrier for the first time, the convertible ceases to exist. Instead, the convertible holder receives the call price. Since the call occurs at the point where call price and conversion value are equal (Ingersoll, 1977a, footnote 17 on p. 305), we may assume that the bondholder converts (contrary to Ingersoll, who makes the equally justified assumption that the bondholder accepts the call price).

This derivation is only valid under Ingersoll's assumptions, especially regarding the growth rate of the barrier. In particular, we need the condition $\rho \leq r$. Using equation (5.7), we can then price the callable convertible discount bond as follows:

$$\begin{aligned}
D_0^{cc} &= \Omega_{uO}(V,T,l_0,\rho) - C_{uO}(V,D,T,l_0,\rho) \\
&\quad + \gamma C_{uO}(V,D/\gamma,T,l_0,\rho) + \gamma\Omega_{uI}(V,T,l_0,\rho) \\
&= \Omega(V,T) - C(V,D,T) + \gamma C_{uO}(V,D/\gamma,T) \\
&\quad - (1-\gamma)\Omega_{uI}(V,T,l_0,\rho) + C_{uI}(V,D,T,l_0,\rho) \\
&\quad - \gamma C_{uI}(V,D/\gamma,T,l_0,\rho),
\end{aligned} \tag{5.9}$$

where $l_0 = De^{-\rho T}/\gamma$, and we have made repeated use of the in-out-parity (cf. equation (3.28)). Our motivation for this was to provide a formula similar to Ingersoll (1977a, equation (19) on p. 307). Comparing equations (5.7) and (5.9), we see that the price of the callable convertible discount bond equals the price of a non-callable convertible, less a discount representing the call feature. Whereas a direct economic interpretation of the discount term in Ingersoll's version of (5.9) is almost impossible without going through each step of the derivation, such an intuitively appealing and justified interpretation is readily available for the formula given here.

Noting that (for $\gamma \leq 1$)

$$\gamma C_{uO}(V, D/\gamma, T, l_0, \rho) = 0$$

because of $l_t \leq D/\gamma$ for $\rho \geq 0$ and $t \leq T$, we can simplify equation (5.9) further to

$$D_0^{cc} = \Omega_{uO}(V, T, l_0, \rho) - C_{uO}(V, D, T, l_0, \rho) + \gamma \Omega_{uI}(V, T, l_0, \rho). \quad (5.10)$$

The value of equity is then given by

$$\begin{aligned} E_0 &= V_0 - D_0^{cc} \\ &= \Omega(V, T, l_0, \rho) - \Omega_{uO}(V, T, l_0, \rho) + C_{uO}(V, D, T, l_0, \rho) \\ &\quad - \gamma \Omega_{uI}(V, T, l_0, \rho) \\ &= (1 - \gamma)\Omega_{uI}(V, T, l_0, \rho) + C_{uO}(V, D, T, l_0, \rho), \end{aligned} \quad (5.11)$$

whereby the equality $V_0 = \Omega(V, T, l_0, \rho)$ holds because $\beta = 0$ (i.e., there are no intermediate payouts to security holders). An economic interpretation of equation (5.10) tells us that, if the up-barrier is not touched during the debt's lifetime, the convertible debt is no different from ordinary debt. In this case, equity corresponds to a call on the firm's assets with strike D. If, however, the barrier is touched, equityholders lose a fraction γ of assets to bondholders. In this case, the requirement to repay the debt immediately ceases to exist.

In the remaining parts of the paper, Ingersoll (1977a) analyzes convertible bonds with continuous coupons and convertible preferred stocks. Furthermore, he studies examples where voluntary (premature) conversion is optimal.

5.5 Finite-Maturity, Discrete-Coupon Bonds, Intermediate Default, Discrete Dividends, Taxes, Stochastic Interest Rates (Brennan and Schwartz 1977,1978,1980)

The work by Brennan and Schwartz (1977) can be viewed as an extension of the Ingersoll (1977a) paper. The main differences are that Brennan and

Schwartz (1977) allow for both discrete coupons and discrete dividends. However, these extensions are not without a cost: Closed-form solutions for debt or equity cannot be derived. Instead, Brennan and Schwartz (1977) recommend a numerical solution algorithm of the finite difference type.

The Brennan and Schwartz (1978) article extends their previous model by introducing tax considerations. Working in the spirit of Modigliani and Miller (1958, 1963), they compare the values of two (otherwise identical) firms that differ only in their capital structure: One firm is purely equity-financed, and the other has a single debt issue outstanding. The value of the unlevered firm between two dividend payments is modelled as a geometric Brownian motion, and the model allows for discrete coupons. The corporation pays taxes at a fixed rate. Interest payments are tax deductible, but dividend payments are not.

The main goal of Brennan and Schwartz (1978) is to investigate the effect of leverage, business risk and payout policy on firm value in this setting. Similar to the Brennan and Schwartz (1977) model, closed-form solutions cannot be derived, and finite difference methods are employed to obtain numerical solutions to the corresponding differential equations.

Brennan and Schwartz (1980) extend their previous work by allowing for stochastic interest rates and senior debt. They model the interest rate development using a mean-reverting stochastic process. As in Brennan and Schwartz (1977, 1978), no closed-form solutions for corporate securities can be derived, and they resort to numerical procedures to solve the respective partial differential equations.

Since the focus here is on models allowing for closed-form solutions, these models (all of which require numerical techniques) are not described in more detail here.

5.6 Warrants (Galai and Schneller 1978)

Galai and Schneller (1978) follow suggestions by Black and Scholes (1973, p. 648 f.) to value warrants as call options on the value of an unlevered firm. A warrant is the right to buy a share of a firm at a certain price during a given time period (or, alternatively, at a certain date in the future), issued by the same firm into whose stocks it is exerciseable. Thus, the main difference between a warrant and a call option is that a warrant is written by the company itself, whereas a call option is written by a third party.

Upon exercise of a warrant, the number of outstanding shares increases. This leads to a *dilution effect*, similar to that described for convertible bonds in Section 5.4: The market value of equity is split among a larger number of shares, which causes the individual share value to decrease. The price received upon warrants issuance compensates shareholders for this possible loss.

Galai and Schneller (1978) assume that the proceeds from warrants issuance are immediately distributed to current shareholders as a cash dividend. Thus, the firm value process (V_t) is left unchanged. Comparing firm values at warrant maturity with and without warrants, they find a model-independent relation between the price of a warrant and that of an ordinary call option.[1] Denoting the ratio of the number of warrants to the number of shares (before warrants issuance) by q (i.e., $q = m/n$ in our notation), they show that

$$W_0 = C_0/(1+q), \tag{5.12}$$

where W_0 denotes the warrant price.

To see this, note that at maturity of the warrants (at time $t = T$), their aggregate value is given by

$$mW_T = \left(\frac{m}{m+n}\right)(V_T - nx)_+,$$

where x denotes the strike price of each warrant. Dividing by m and taking the expectation under $\mathbb{Q}^B$, we get the time $t = 0$ price of one warrant:

$$W_0 = \frac{1}{m+n}\mathbb{E}^B[e^{-rT}(V_T - nx)_+].$$

Using the well-known general relation $C(V, K, T) = aC(V/a, K/a, T)$ (with $a \in \mathbb{R}_+$), we get further

$$W_0 = \frac{n}{m+n}C\left(\frac{V_0}{n}, x, T\right).$$

Noting that

$$\frac{1}{1+q} = \frac{n}{m+n},$$

[1] Galai and Schneller (1978) argue initially within a one-period model. Therefore, validity of this relation is only assured for European warrants and options. In fact, the relation is also valid for all cases where premature exercise is non-optimal (see, e.g., Emanuel (1983), Constantinides (1984) and Spatt and Sterbenz (1988)).

and, by the assumption of a purely equity-financed firm (i.e., $V_0/n = S_0$), equation (5.12) follows.

Alternative assumptions regarding the use of the proceeds from warrants issuance lead to different valuation formulae for the warrants. E.g., assuming that the proceeds from warrants issuance lead to a level shift in the asset value process (but do not otherwise change its distribution), we obtain the standard warrant valuation equation found in textbooks (e.g. in Hull (1997, pp. 245f.); for a discussion of this, see Hanke and Pötzelberger (2002, pp. 67f.)).

5.7 Empirical Study of Firm Value Based Pricing of Corporate Bonds (Jones, Mason and Rosenfeld 1984)

Jones, Mason, and Rosenfeld (1984) conduct an empirical analysis of firm value based pricing of corporate bonds. Starting from the standard partial differential equation for callable debt already derived by Merton (1974), they perform a numerical analysis to find equity-maximizing call policies and resulting bond prices. They test their model empirically on a sample of 27 firms for data from the late 1970s. Since their approach does not lead to closed-form solutions for prices of corporate securities, a more detailed description of the model is omitted.

As far as the empirical results are concerned, the authors observe that their model fails to generate yield spreads at levels as high as observed in practice. They conclude that the explanatory power of firm value based pricing models could be enhanced by introducing taxes and stochastic interest rates.[2] However, there are a number of problems with their empirical analysis, which were pointed out in a discussion of the paper by Fisher (1984). His comments make the empirical findings of Jones, Mason, and Rosenfeld (1984) questionable, and we refrain from a more detailed discussion here.

---

[2]Note that Kim, Ramaswamy, and Sundaresan (1993) later found that yield *spreads* are quite insensitive to interest rate uncertainty.

5.8 Finite-Maturity, Continuous-Coupon Bonds, Intermediate Default, CIR Interest Rates (Kim, Ramaswamy and Sundaresan 1993)

Kim, Ramaswamy, and Sundaresan (1993) extend the structural (or firm value based) security pricing approach to allow for stochastic interest rates. Their approach can be viewed as a combination of the Black and Cox (1976) bond pricing model and the Cox, Ingersoll, and Ross (1985) interest rate model. The short rate process in this model is mean-reverting and given by

$$dr = \kappa(\overline{\mu} - r)dt + \sigma_r \sqrt{r}\, d\overline{W}_t,$$

where $\overline{\mu}$ is the long-run mean value of the short rate and $(\overline{W}_t)$ is a Wiener process different from (but possibly correlated with) that driving the firm value. Since this extension renders the model analytically intractable (i.e., precludes the derivation of closed-form solutions), this model is of less interest to us. The numerical results, however, are encouraging: Kim, Ramaswamy, and Sundaresan (1993) find that their model generates yield spreads of the magnitude observed in practice. Moreover, they find that (p. 118)

> "...yield spreads between [non-callable] Treasury and corporate bonds are quite insensitive to interest rate uncertainty."

This provides a justification for using firm value based pricing models with the assumption of constant interest rates to price, e.g., credit derivatives based on spreads.

It is important to note that they also find that

> "stochastic interest rates seem to play an important role in determining the yield differentials between a callable corporate bond and an equivalent government bond due to the interactions between call provisions and default risk."

This means that firm value based pricing models with the assumption of constant interest rates (featuring closed-form solutions) may be less suitable for pricing credit derivatives based on spreads of callable corporate bonds.

5.9 Finite-Maturity, Continuous-Coupon Bonds, Intermediate Default, Vasicek Interest Rates (Longstaff and Schwartz 1995)

Similar to Kim, Ramaswamy, and Sundaresan (1993), Longstaff and Schwartz (1995) combine "traditional" firm value based security pricing with stochastic interest rates. However, they choose a simpler framework for interest rate uncertainty originally proposed by Vasicek (1977). In this model, the short rate is described by the following stochastic process:

$$dr = (\overline{\mu} - r)dt + \sigma_r d\overline{W}_t.$$

Sacrificing flexibility in interest rate modelling[3] pays off in terms of quasi-closed form solutions (an integral equation that can be approximated numerically) for the prices of fixed and floating rate bonds.

Since up to now, no option pricing extensions of this model exist, we do not describe it in more detail here. As far as the empirical results are concerned, it is interesting to note their finding that (Longstaff and Schwartz, 1995, p. 812)

> "these results [...] provide clear evidence against the traditional approach to valuing risky debt in which the interest rate is assumed to be constant [...]. In fact, the variation in credit spreads due to changes in the level of interest rates is more important for these investment-grade bonds than the variation due to changes in the value of the firm."

This is in sharp contrast to the findings of Kim, Ramaswamy, and Sundaresan (1993). Whereas the valuation of equity and derivatives on equity, which is the topic of this book, may be less sensitive to interest rate uncertainty, the results of Longstaff and Schwartz (1995) suggest that one should be cautious when applying firm value based models assuming deterministic risk-free interest rates for valuing debt and credit derivatives, in particular for firms with low credit risk.

---

[3] For a brief discussion of possible disadvantages of the Vasicek model together with a justification of using it in this context, see Longstaff and Schwartz (1995, p. 792).

5.10 Infinite-Maturity, Continuous-Coupon Bonds, Taxes, Intermediate Default, Bankruptcy Costs (Leland 1994)

Leland (1994) also works in an extended Black–Scholes framework with total firm value modelled as a geometric Brownian motion. In addition, his model allows for taxes,[4] bankruptcy costs and protective covenants. Using the partial differential equation approach, he finds that closed-form solutions for equity and debt in this setting can be derived if security payoffs are time-independent. This time-independence condition is satisfied, e.g., if securities have infinite maturities (conditional on no prior default). Whereas this assumption is certainly justified for stocks, corporate bonds usually have finite maturities. Leland justifies his assumption (apart from mathematical convenience) by noting that in the case of very long term debt contracts, repayment of principal constitutes only a small percentage of debt value (Leland, 1994, p. 1215). Moreover, in this way many small debt issues can be conveniently modelled collectively as one large issue with continuous coupon payments.

One limitation of Leland's model is a restriction on the barrier level, which is not made in a very explicit way and is not discussed any further. It is only when bankruptcy costs are introduced that Leland touches this question briefly, stating that (the notation has been adapted)

> "If bankruptcy occurs, a fraction $0 \leq \varphi^K \leq 1$ of value will be lost to bankruptcy costs, leaving debtholders with value $(1-\varphi^K)V_\tau$ and stockholders with nothing."

Using this as a basis for his boundary conditions, Leland effectively rules out barriers $L > D_0/(1-\varphi^K)$. For barriers exceeding this threshold, the boundary condition (6i) in Leland (1994, p. 1219):

$$\text{“At } V = L = V_\tau, \quad D(V) = (1-\varphi^K)V_\tau\text{”}$$

would not be economically sensible, since it would imply that in case of bankruptcy, bondholders receive more than the value of debt at the time

[4] Fischer, Heinkel, and Zechner (1989) is another paper using firm value as the underlying state variable in the presence of taxes.

of bankruptcy.[5] However, such barrier levels might be economically interesting, leading to cases where equity receives positive payments in the case of default.

Given these restrictions, as far as equity is concerned, Leland's model can be viewed as a special case of the Ericsson and Reneby (1998) model with infinite model horizon, constant barrier, and only one bond outstanding (which pays a continuous coupon), with the additional restriction regarding the barrier just discussed. This holds because, with this restriction, there are no payoffs to equity in case of bankruptcy.

Lifting the restriction regarding the barrier, and imposing instead the assumptions of Ericsson and Reneby (1998) regarding the distribution of asset value between security holders in case of bankruptcy, we can write equity as a down-and-out asset claim with infinite maturity less the value of a down-and-out continuous coupon stream (corrected for the tax rate) plus φ^E down-and-in asset claims:

$$E(\cdot,\infty) = \Omega_{LO}(V,\infty,L) - (1-\zeta)C \cdot U_{LO}(V,\infty,L) + \varphi^E \Omega_{LI}(V,\infty,L), \quad (5.13)$$

where C denotes the coupon rate. Equation (5.13) is the extended version of the equity valuation equation in the Leland (1994) model, allowing for barrier levels $0 \leq L < V_0$.

To show the correspondence to Leland's result, impose his restriction again (i.e., set $\varphi^E = 0$) and substitute into equation (5.13) to get

$$E(\cdot,\infty) = V - \left(\frac{L}{V}\right)^{\alpha} \frac{L^2}{V} - (1-\zeta)\frac{C}{r}\left[1 - \left(\frac{V}{L}\right)^{-\theta}\right], \quad (5.14)$$

where we have made use of equations (4.6), (3.27), (3.28) and (4.8), α is defined in equation (3.5), and θ is defined in equation (4.4).

Rewriting equation (13) in Leland (1994) using our symbols, it reads as

$$E(\cdot,\infty) = V - (1-\zeta)\frac{C}{r} + \left[(1-\zeta)\frac{C}{r} - L\right]\left(\frac{V}{L}\right)^{-\frac{2r}{\sigma^2}}. \quad (5.15)$$

Comparing equations (5.14) and (5.15) and cancelling obviously identical terms (note that θ collapses to $\frac{2r}{\sigma^2}$), it remains to show that

$$\left(\frac{L}{V}\right)^{\alpha} \frac{L}{V} = \left(\frac{V}{L}\right)^{-\frac{2r}{\sigma^2}}.$$

[5] As discussed by Longstaff and Schwartz (1995, p. 794), this situation may occur in rare circumstances, but mostly for cases ruled out here (non-constant interest rates).

Thus, the question is whether $1 + \alpha = \frac{2r}{\sigma^2}$. Recalling from equation (3.5) that $\alpha = \frac{2(r-\sigma^2/2)}{\sigma^2}$ (if, as is the case here, $\beta = 0$), this identity follows immediately.

Leland goes on to analyze different variants for the determination of the barrier on asset value and distinguishes two cases: Unprotected debt, where the optimal (in terms of maximum equity value) bankruptcy threshold is determined endogenously, and protected debt, where the bankruptcy threshold is determined by debt covenants. For each case, he determines firm value-maximizing leverage levels and analyzes their dependencies on the model parameters. In particular, he derives the equity-maximizing endogenous bankruptcy level as (Leland, 1994, p. 1222)

$$L_{end} = \frac{(1-\zeta)C}{r(1+\frac{\sigma^2}{2r})}. \tag{5.16}$$

Comparing equations (5.16) and (5.6), it is easy to see that the difference merely lies in the extension for taxes.

In a later section of his paper, Leland (1994, p. 1241) extends his model to allow for net cash outflows proportional to V. This extended model forms the basis for the option pricing extension provided by Toft and Prucyk (1997) (discussed in Section 7.2). He assumes that the process for (V_t) under $\mathbb{P}$ is given by

$$dV_t = (\mu(V_t, t) - \beta)V_t dt + \sigma V_t dW_t.$$

In this case, the equity valuation formula in equation (5.13) has to be extended for the "dividend payments" to equity, which correspond to β down-and-out asset streams (assuming that $\varphi^E = 0$):

$$E(\cdot, \infty) = \Omega_{LO}(V, \infty, L) + \beta O_{LO}(V, \infty, L) - (1-\zeta)C \cdot U_{LO}(V, \infty, L). \tag{5.17}$$

This extension requires an extension to the formula for the endogenous barrier:

$$L_{end} = \frac{(1-\zeta)C\theta}{r(1+\theta)}. \tag{5.18}$$

Substituting for θ in equation (5.18), it is easy to verify that this equation collapses to equation (5.16) in the no-payout case ($\beta = 0$).

Leland continues by examining the dependence of debt values and yield spreads on leverage, firm risk, taxes, covenants and bankruptcy costs, and finds that junk bonds behave in many respects qualitatively differently from bonds with low credit risk. Finally, he focuses on the role of protective covenants in reducing agency problems between debt and equity and takes a look at the effects of debt renegotiation.

5.11 Finite-Maturity, Continuous-Coupon Bonds, Taxes, Intermediate Default, Bankruptcy Costs (Leland and Toft 1996)

Leland and Toft (1996) extend the Leland (1994) model to allow for finite-maturity debt. To enable them to work in a stationary capital structure setting as in Leland (1994), they assume that the firm continuously sells a constant amount of new debt (principal) with given (and constant) maturity T and coupon rate C. This makes total debt service payments time-independent, which was an important condition in the derivation of securities prices via partial differential equations in Leland (1994).

At every instant of time, total outstanding principal is a constant D. To determine the value of debt currently outstanding, note that its "payoffs in solvency" consist of two parts: a constant stream of principal repayment within the time interval $[0,T]$, and C linearly decreasing coupon streams (where the linear decrease in coupons is due to the constant repayment of principal):

$$\Gamma(D^S) = \frac{D}{T}\Gamma(\mathrm{U}(V,T)) + C\Gamma({}^{\mathrm{dec}}_{\mathrm{lin}}\mathrm{U}(V,T)). \tag{5.19}$$

As in the Leland (1994) model, in case of default, a fraction φ^K of asset value is lost, and the remainder $(1-\varphi^K)$ is distributed to bondholders. This has similar implications with regard to admissible barrier levels as discussed in the previous section. Assuming equal seniority of all outstanding debt issues, this means that the payoff to currently outstanding debt in case of default is given by

$$\Gamma(D^\tau) = (1-\varphi^K)V_\tau. \tag{5.20}$$

However, each outstanding issue is entitled to only a fraction of this payoff. Equal seniority implies that these fractions add up to $\varphi^D = 1-\varphi^K$.

Adding up the components in equations (5.19) and (5.20) and substituting prices for payoffs, we get

$$\begin{aligned} D(V,T,L,C) = &\frac{D}{T} U_{LO}(V,T,L) + C \cdot {}^{dec}_{lin}U_{LO}(V,T,L) \\ &+ \varphi^D L \cdot {}^{dec}_{lin}G_{LI}(V,T,L). \end{aligned} \tag{5.21}$$

To derive the value of equity when debt with maturity T is issued according to the scheme described above, we follow Leland and Toft (1996, p. 992) again and note that the value of the firm's securities is equal to the asset value plus tax benefits from coupons less the value of the claim to bankruptcy costs. The aggregate value of the firm's securities is given by (see Leland and Toft (1996, equation 8))

$$E(\cdot) + D(\cdot) = V + \zeta C \cdot U_{LO}(V,\infty,L) - \varphi^K L \cdot G_{LI}(V,\infty,L). \tag{5.22}$$

From equation (5.22), the value of equity can be calculated using the expression for debt given in equation (5.21):

$$\begin{aligned} E(\cdot) = &V + \zeta C \cdot U_{LO}(V,\infty,L) - \varphi^K L \cdot G_{LI}(V,\infty,L) \\ &- \frac{D}{T} U_{LO}(V,T,L) - C \cdot {}^{dec}_{lin}U_{LO}(V,T,L) \\ &- \varphi^D L \cdot {}^{dec}_{lin}G_{LI}(V,T,L). \end{aligned} \tag{5.23}$$

Remark 5.1. At first sight, one could be tempted to think that future debtholders have been left out of the analysis. However, note that under the assumption that future debt issues will be floated at their fair values, the discounted expected values (under $\mathbb{Q}^B$) of debt service payments will equal the issuing price. Therefore, the net effect consists only of the tax benefits, which have been included in equation (5.22). The restrictive assumption regarding the barrier could easily be overcome by lifting the requirement that $\varphi^D = (1 - \varphi^K)$. If $\varphi^K + \varphi^D < 1$, the difference is the fraction of asset value that goes to equity in case of default.

Instead of using an externally given bankruptcy trigger, Leland and Toft (1996, p. 993) determine an endogenous, equity-maximizing threshold L_{end} similarly to Black and Cox (1976) and Leland (1994) (rewritten here in our notation):

$$L_{end} = \frac{C\left(\frac{A}{rT} - B\right) - \frac{A \cdot D}{T} - \frac{\zeta C(\mu_X^B - \mu_X^G)}{\sigma}}{r\left(1 + \frac{\varphi^K(\mu_X^B - \mu_X^G)}{\sigma} - \varphi^D B\right)},$$

where

$$A = \frac{1}{\sigma}\left(2\mu_X^B e^{-rT} N(\mu_X^B\sqrt{T}) + 2\mu_X^G N(-\mu_X^G\sqrt{T}) - \frac{2}{\sqrt{T}} n(-\mu_X^G\sqrt{T}) \right.$$
$$\left. + \frac{2e^{-rT}}{\sqrt{T}} n(\mu_X^B\sqrt{T}) - \mu_X^B - \mu_X^G\right),$$
$$B = \frac{1}{\sigma}\left(\left(2\mu_X^G + \frac{2}{\mu_X^G T}\right) N(-\mu_X^G\sqrt{T}) - \frac{2}{\sqrt{T}} n(-\mu_X^G\sqrt{T}) \right.$$
$$\left. - \mu_X^B - \mu_X^G - \frac{1}{\mu_X^G T}\right),$$

and $n(\cdot)$ denotes the standard normal density function. In the limiting case of $T \to \infty$, this endogenous bankruptcy trigger turns out to be the same as in the Leland (1994) model (cf. equation (5.16)). Thus, bankruptcy can be modelled as a discretionary decision by equityholders. The authors analyze the implications of their approach and find that debt maturity has a major influence on this endogenously determined bankruptcy level: Whereas for long-term debt, bankruptcy triggers will be below debt principal, they find that for short-term debt, rational equityholders will declare bankruptcy at asset levels that exceed debt principal.

The authors continue with an analysis of optimal leverage values, debt capacities depending on leverage, term structures of credit spreads and bankruptcy rates. In general, Leland and Toft (1996) find that their model generates (optimal) leverage values, credit spreads, default rates and write-downs that are close to historical averages. An interesting finding is that the "true duration" of high-risk bonds (taking default risk properly into account, as opposed to Macaulay duration) is not only smaller than the Macaulay duration, but may even become negative.

In the remainder of the paper, Leland and Toft (1996) study the relation between agency effects (in particular, asset substitution) and debt maturity. In addition, they discuss how multiple classes and seniorities of debt could be modelled by adjusting the fraction of firm value received by each class in case of bankruptcy accordingly.

5.12 Finite-Average-Maturity, Continuous-Coupon Bonds, Taxes, Intermediate Default, Bankruptcy Costs, Costly Debt Issuance (Leland 1998)

In this article, Leland extends his previous work in various ways. First, instead of infinite-maturity, continuous-coupon debt (Leland, 1994) or a linear repayment schedule (Leland and Toft, 1996), he uses an exponential debt repayment schedule. He assumes that, initially, debt is issued with principal $D = D^{in}$ and a coupon rate of C per year. At any time $t > 0$, a fraction $mD^{in}(t)$ will be retired, where $D^{in}(t)$ denotes initial principal still outstanding at time t. Thus, at any time $t > 0$, e^{-mt} of debt issued at time $t = 0$ will remain outstanding, with principal $e^{-mt}D^{in}$ and coupon payment rate $e^{-mt}C$. As noted by Leland (1998, p. 1218), this leads to an average debt maturity $M = 1/m$:

$$M = \int_0^\infty t \, \frac{me^{-mt}D}{D} dt = \frac{1}{m}. \tag{5.24}$$

Assuming that retired debt is continuously replaced by new debt with the same principal, coupon rate and seniority, this leads to a constant capital structure conditional on no restructuring. Debt issuance costs are explicitly taken into account: Initial debt issuance incurs costs of $k_1 D$, whereas newly issued principal costs $k_2 mD$.

Restructuring may occur in two different ways: If (V_t) reaches some lower bound L, default occurs (in the sense already discussed for the Leland (1994) and Leland and Toft (1996) models), and the company's assets (after subtracting bankruptcy costs) are divided among claimants according to pre-specified rules. If, however, (V_t) reaches some upper bound U without prior touching of L, all debt will be retired at par and new debt will be issued. D, C, L and U will be scaled up by the increase of asset value, U/V_0. This may happen, in theory, infinitely often.

Leland (1998) derives the value of debt and equity in closed form. The equity-maximizing optimal barrier has to be solved for using numerical techniques. Since an option pricing extension of this economically extremely rich capital structure model proves to be a daunting task,[6] we will confine ourselves in Chapter 8 to derive an option pricing formula for a restricted

[6] One way to approach this would be via an extension of the work of Geman and Yor (1996).

version of this model that does not allow for restructurings at some upper boundary.

The value of initially issued debt corresponds to $(C+mD)$ exponentially decreasing down-and-out unit streams:

$$D_0(\cdot) = (C + mD)U_{LO}^{-m}(V, \infty, L) + \varphi^D L \cdot G_{LI}^{-m}(V, \infty, L). \tag{5.25}$$

Assuming that each new debt issue is serviced according to the same exponential repayment schedule, debt issuance costs correspond to $k_2 mD$ down-and-out unit streams of infinite maturity. Total firm value is given by the sum of asset value and tax benefits less the value of bankruptcy costs and costs of debt issuance:

$$\begin{aligned} E_0(\cdot) + D_0(\cdot) = V + \zeta C \cdot U_{LO}(V, \infty, L) - \varphi^K L \cdot G_{LI}(V, \infty, L) \\ - k_1 D - k_2 mD \cdot U_{LO}(V, \infty, L). \end{aligned} \tag{5.26}$$

Using the result for debt value in equation (5.25), the value of equity can be calculated from equation (5.26) as

$$\begin{aligned} E_0(\cdot) = V + \zeta C \cdot U_{LO}(V, \infty, L) - \varphi^K L \cdot G_{LI}(V, \infty, L) - k_1 D \\ - k_2 mD \cdot U_{LO}(V, \infty, L) - (C + mD)U_{LO}^{-m}(V, \infty, L) \\ - \varphi^D L \cdot G_{LI}^{-m}(V, \infty, L). \end{aligned} \tag{5.27}$$

The focus in Leland (1998) is on capital structure decisions which are optimal from the viewpoint of equity together with their implications for other claimants. He notes that, in contrast to much of the previous capital structure literature (that focused on qualitative results), this model

> "provides quantitative guidance on the amount and maturity of debt, on financial restructuring, and on the firm's optimal risk strategy." (p. 1237)

5.13 "Model A": Finite-Maturity, Continuous-Coupon Bonds, Exponentially Increasing Debt, Intermediate Default, Bankruptcy Costs, Taxes, Deviations from Absolute Priority (Extended Leland and Toft)

One of the main arguments of Reneby (1998) and Ericsson and Reneby (2001) to favour exponentially increasing debt over constant debt was that firms are usually expected to grow, i.e., the drift of (V_t) is positive. If debt were not allowed to grow, leverage would gradually decline to zero, which is clearly unrealistic. Moreover, Ericsson and Reneby (2001, pp. 8f.) show that a constant barrier leads to unrealistic default intensities: If the firm does not go bankrupt in the near future, it never will.

For these reasons, they propose a model where both total debt and the barrier grow exponentially. However, their model only features infinite maturity debt.[7] In order to study effects of debt maturity, an extension to finite maturity debt would be desirable.

The model we propose in this section (henceforth referred to as "Model A") can be viewed as such an extension. Alternatively, it can be viewed as an extension of the Leland and Toft (1996) model to the case of exponentially increasing debt. We make the following assumptions:

- The initial debt structure is similar to that in the Leland and Toft (1996) model, i.e., bond maturities are uniformly distributed in the interval $[0, T]$. The main difference is that the principal of the individual bond issues is exponentially increasing in time (at rate ν). E.g., the bond that will mature in the next time instant has a principal of $(D/T)dt$, whereas the bond with the longest maturity has a principal of $(e^{\nu T}D/T)dt$ (analogously for the coupon). Coupons are a constant fraction of principal, and therefore also geometrically increasing at rate ν.

- At every instant of time, currently maturing debt is replaced by new debt with maturity T and a principal of $e^{\nu T}$ times the principal of the retired debt. The new debt is floated at market value.

- The barrier is assumed to grow exponentially at rate ρ. If $\rho = \nu$, the recovery ratio for debt remains constant over time, so this seems to be a plausible choice for ρ from an economic point of view.

- φ^D, the fraction of asset value paid to debtholders in case of default, is constant. If $\rho = \nu$, this means that the "additional protection"

[7] More precisely, *total* debt is assumed to be of infinite maturity. The main purpose of the Ericsson and Reneby (2001) model is to price small single debt issues relative to equity and total debt. Since they assume that a single debt issue is too small to have any effect on equity, their model is clearly not suitable for investigating effects of debt maturity on option prices.

offered by the increasing barrier serves to keep the recovery ratio for total debt in case of default constant.

Under these assumptions, the value of initially outstanding debt is given by (cf. equation (5.21))

$$\begin{aligned} D(\cdot) = & \frac{D}{T} U_{lO}^{\gamma}(V, T, l_0, \rho) + C \cdot {}^{\text{dec}}_{\text{lin}} U_{lO}^{\gamma}(V, T, l_0, \rho) \\ & + \varphi^D l_0 \cdot {}^{\text{dec}}_{\text{lin}} G_{lI}^{\gamma}(V, T, l_0, \rho). \end{aligned} \tag{5.28}$$

Total firm value (cf. equation (5.22)) will reflect our assumptions regarding the growth rate of debt through increased tax benefits in the future.

$$E(\cdot) + D(\cdot) = V + \zeta C \cdot U_{lO}^{\gamma}(V, \infty, l_0, \rho) - \varphi^K l_0 \cdot G_{lI}^{\rho}(V, \infty, l_0, \rho). \tag{5.29}$$

From equation (5.29), the value of equity can be calculated using the value of debt given in (5.28):

$$\begin{aligned} E(\cdot) = & V + \zeta C \cdot U_{lO}^{\gamma}(V, \infty, l_0, \rho) - \varphi^K l_0 \cdot G_{lI}^{\rho}(V, \infty, l_0, \rho) \\ & - \frac{D}{T} U_{lO}^{\gamma}(V, T, l_0, \rho) - C \cdot {}^{\text{dec}}_{\text{lin}} U_{lO}^{\gamma}(V, T, l_0, \rho) \\ & - \varphi^D l_0 \cdot {}^{\text{dec}}_{\text{lin}} G_{lI}^{\gamma}(V, T, l_0, \rho). \end{aligned} \tag{5.30}$$

Chapter 6

Extension of the Probabilistic Security Pricing Framework to Derivative Securities

In this chapter, we review an extension to the framework presented in Chapter 4 due to Ericsson and Reneby (1996). This extension enables us to value options on corporate securities. By deriving valuation formulae for an array of additional claims, we will extend this option pricing framework considerably. We will show that existing firm value based option pricing models can be conveniently expressed within this extended framework. In Chapters 7 and 8, we will use it to derive option pricing formulae for the capital structure models described in Chapter 5.

6.1 Ericsson and Reneby (1996)

6.1.1 Assumptions and Results

Ericsson and Reneby (1996) extend their 1998 work described in Section 4.1 (which provided a framework for valuing corporate securities in the presence of a constant barrier on firm value) to the valuation of options on general corporate securities. To this end, they first extend the Geske (1979) model to the case where the underlying call option is down-and-out. They provide a new interpretation for the parts of the compound option pricing formula and introduce the notion of *conditional options*. Then, they show how options on general corporate securities (as in Proposition 4.1) can be priced as sums of conditional calls and heavisides minus a "regular" down-and-out heaviside representing the strike price.

The general setup with regard to the firm value process and the barrier is the same as in Ericsson and Reneby (1998). Starting with the down-and-out call option on V with strike F and maturity T_2 (whose price was given in Lemma 4.3), their first step is to derive the price of a call option on this down-and-out call with strike K and maturity $T_1 < T_2$. Define the events

- $A_{T_1} = \{V_{T_1} > \overline{V}, \tau \notin [0, T_1]\}$ and
- $A_{T_2} = \{V_{T_2} > F, V_{T_1} > \overline{V}, \tau \notin [0, T_2]\}$.

Assuming that $\overline{V}, F, V > L$ (parts of this assumption will be criticized and dropped in Section 6.1.2), Ericsson and Reneby (1996) derive the following probabilities (under different measures) for the event A_{T_2}:

Lemma 6.1. *The probabilities of the event* A_{T_2} *under the measures* $\mathbb{Q}^m$ *with* $m \in \{V, B\}$ *are given by*

$$\begin{aligned}\mathbb{Q}^m\{A_{T_2}\} = {} & N\left(d^m_{T_1}\left(\frac{V}{\overline{V}}\right), d^m_{T_2}\left(\frac{V}{F}\right), \sqrt{\frac{T_1}{T_2}}\right) \\ & - \left(\frac{L}{V}\right)^{-\frac{2}{\sigma}\cdot\mu^m_X} N\left(d^m_{T_1}\left(\frac{L^2}{V\cdot\overline{V}}\right), d^m_{T_2}\left(\frac{L^2}{V\cdot F}\right), \sqrt{\frac{T_1}{T_2}}\right) \\ & - \left(\frac{L}{V}\right)^{-\frac{2}{\sigma}\cdot\mu^m_X} N\left(-d^m_{T_1}\left(\frac{\overline{V}}{V}\right), d^m_{T_2}\left(\frac{L^2}{V\cdot F}\right), -\sqrt{\frac{T_1}{T_2}}\right) \\ & + N\left(-d^m_{T_1}\left(\frac{V\cdot\overline{V}}{L^2}\right), d^m_{T_2}\left(\frac{V}{F}\right), -\sqrt{\frac{T_1}{T_2}}\right)\end{aligned} \tag{6.1}$$

where d^m_t *and* μ^m_X *are defined as in equations* (4.2) *and* (4.3).

The probabilities of the event A_{T_1} under the various probability measures have already been given in Lemma 4.1.

Lemma 6.2. *The price of a call with exercise price* K *and maturity* T_1 *on a down-and-out call with strike* F *of maturity* T_2 *is given by*

$$\begin{aligned}& C(C_{LO}(V, F, T_2, L), K, T_1) = \\ & \qquad V\cdot\mathbb{Q}^V(A_{T_2}) - e^{-rT_2}F\cdot\mathbb{Q}^B\{A_{T_2}\} - e^{-rT_1}K\cdot\mathbb{Q}^B\{A_{T_1}\},\end{aligned} \tag{6.2}$$

where $\overline{V}$ *in* A_{T_1}, A_{T_2} *is defined by*

$$\overline{V} := C_{LO}(\overline{V}, F, T_2 - T_1, L) = K. \tag{6.3}$$

Further, Ericsson and Reneby (1996) define *conditional* options as derivatives with the usual payoffs conditional on an event A. In particular, they define a conditional down-and-out call option as

$$\Gamma(\mathcal{C}_{LO}(V, F, T_2, L|A)) = \Gamma\left(C_{LO}(V, F, T_2, L)\right) \cdot I_{\{A\}},$$

and a conditional down-and-out heaviside option as

$$\Gamma(\mathcal{H}_{LO}(V, F, T_2, L|A)) = \Gamma\left(H_{LO}(V, F, T_2, L)\right) \cdot I_{\{A\}}.$$

From their derivation of Lemma 6.2, it is obvious that the prices of these claims for the case $V_{T_1} > \overline{V}$ are given by

$$\mathcal{C}_{LO}(V, F, T_2, L|\{V_{T_1} > \overline{V}\}) = V \cdot \mathbb{Q}^V(A_{T_2}) - e^{-rT_2}F \cdot \mathbb{Q}^B\{A_{T_2}\}$$

and

$$\mathcal{H}_{LO}(V, F, T_2, L|\{V_{T_1} > \overline{V}\}) = e^{-rT_2} \cdot \mathbb{Q}^B\{A_{T_2}\}.$$

Thus, the compound down-and-out call can be interpreted as a conditional call corresponding to the first two terms of equation (6.2) (conditional upon $V_{T_1} > \overline{V}$) less the requirement to pay the strike price if the option turns out to be in-the-money at time $t = T_1$, represented by the last term in equation (6.2):

$$\begin{aligned} C(C_{LO}(V, F, T_2, L), K, T_1) = {} & \mathcal{C}_{LO}(V, F, T_2, L|\{V_{T_1} > \overline{V}\}) \\ & - K \cdot H_{LO}(V, \overline{V}, T_1, L) \end{aligned} \tag{6.4}$$

For purposes of valuing derivatives, Ericsson and Reneby (1996) assume that corporate securities have a post-default value of 0. The central result in Ericsson and Reneby (1996) is the following proposition, which is essentially an extension of the principle underlying the representation of the down-and-out call in equation (6.4):

Proposition 6.1 (Ericsson/Reneby (1996), incorrect!). *If default of a corporate security* CS *as defined in Proposition 4.1 is triggered by at most*

- *a barrier* L *and*
- *the inability to repay debt at maturity* T_2,

then its value can be expressed as

$$
\begin{aligned}
CS(\cdot) = {} & a \cdot V \\
& + \sum_{T_1 < T \le T_2} \sum_i b_{T,i} \cdot C_{LO}(V, F_{T,i}, T, L) \\
& + \sum_{T_1 < T \le T_2} \sum_i c_{T,i} \cdot H_{LO}(V, F_{T,i}, T, L)
\end{aligned}
$$

for $t \le \tau$ *(where* $a, b_{(\cdot)}, c_{(\cdot)}$ *are constants denoting the numbers/fractions of the respective claims), and the time* $t = 0$ *price of a call option of maturity* T_1 *and strike* K *on this corporate security is given by*

$$
\begin{aligned}
C(CS(V, T_2, \cdot), K, T_1) = {} & a \cdot \mathcal{C}_{LO}(V, 0, T_2, L) | \{V_{T_1} > \overline{V}\}) \\
& + \sum_{T_1 < T \le T_2} \sum_i b_{T,i} \cdot \mathcal{C}_{LO}(V, F_{T,i}, T, L | \{V_{T_1} > \overline{V}\}) \\
& + \sum_{T_1 < T \le T_2} \sum_i c_{T,i} \cdot \mathcal{H}_{LO}(V, F_{T,i}, T, L | \{V_{T_1} > \overline{V}\}) \\
& - K \cdot H_{LO}(V, \overline{V}, T_1, L),
\end{aligned}
$$

where $\overline{V}$ *is defined by*

$$
CS_{T_1}(\overline{V}, \cdot) = K.
$$

Remark 6.1. Comparing the first part in Proposition 6.1 to Proposition 4.1, we note that a different representation has been chosen here. The motivation for this is unclear, maybe the fact that many firm value based pricing models found in the literature use formulae containing V has played a role.

Proposition 6.1 is incorrect. This will be shown in the next section, where we will also give a corrected and extended version.[1]

6.1.2 Correcting the Ericsson–Reneby (1996) Results

In order to show where the error in Proposition 6.1 comes from, we restate here parts of the derivation given in Ericsson and Reneby (1996, Appendix C).

[1]In his Ph.D. thesis, Reneby (1998, p. 37) provides yet another formulation, allowing for an exponential barrier and containing unit down-and-in claims. His version is correct, but the problem discussed here does not become obvious in his formulation. Since equity formulae based on V are very common in the literature, we decided to correct and extend the original formulation.

The assumption that corporate securities have a post-default value of 0 (at least for the purpose of pricing derivatives) can be formalized by using $\widehat{CS}$ instead of CS as the option's underlying, where

$$\widehat{CS}(V, T_2, \cdot) = CS(V, T_2, \cdot) \cdot I_{\{\tau \notin [0,T_1]\}}.$$

Applying the martingale valuation principle and inserting, they get

$$C(\widehat{CS}(V, T_2, \cdot), K, T_1) = \begin{cases} a \cdot e^{-rT_1}\mathbb{E}^B[V_{T_1} \cdot I_{\{V_{T_1} > \overline{V}\}} I_{\{\tau \notin [0,T_1]\}}] \\ + \sum_{T_1 < T \leq T_2} \sum_i b_{T,i} \cdot e^{-rT_1}\mathbb{E}^B[C_{LO}(\cdot)|\{V_{T_1} > \overline{V}\}] \\ + \sum_{T_1 < T \leq T_2} \sum_i c_{T,i} \cdot e^{-rT_1}\mathbb{E}^B[H_{LO}(\cdot)|\{V_{T_1} > \overline{V}\}] \\ - K \cdot H_{LO}(V, \overline{V}, T_1, L) \end{cases}. \tag{6.5}$$

Then, they state that "noting that a claim on the assets at T_1 is equivalent to a call with exercise price zero on the assets at T_2 we finally can write

$$C(\widehat{CS}(V, T_2, \cdot), K, T_1) = \begin{cases} a \cdot \mathcal{C}_{LO}(V, 0, T_2, L|\{V_{T_1} > \overline{V}\}) \\ + \sum_{T_1 < T \leq T_2} \sum_i b_{T,i} \cdot \mathcal{C}_{LO}(\cdot|\{V_{T_1} > \overline{V}\}) \\ + \sum_{T_1 < T \leq T_2} \sum_i c_{T,i} \cdot \mathcal{H}_{LO}(\cdot|\{V_{T_1} > \overline{V}\}) \\ - K \cdot H_{LO}(V, \overline{V}, T_1, L) \end{cases} ." \tag{6.6}$$

This is where the error comes in. Denoting the first passage time of (V_t) to L in the interval $[T_1, T_2]$ with τ_{T_2}, the down-and-out call substituted in the first line of equation (6.6) can be written as

$$\mathcal{C}_{LO}(V, 0, T_2, L|\{V_{T_1} > \overline{V}\}) = C(V, 0, T_2) \cdot I_{\{V_{T_1} > \overline{V}\}} \cdot I_{\{\tau \notin [0,T_1]\}} \cdot I_{\{\tau_{T_2} \notin [T_1,T_2]\}}. \tag{6.7}$$

The first line of equation (6.5), however, can be rewritten as (dropping the constant a)

$$\begin{aligned} e^{-rT_1}\mathbb{E}^B[V_{T_1} \cdot I_{\{V_{T_1} > \overline{V}\}}] &= \Omega_{LO}(V, T_1) \cdot I_{\{V_{T_1} > \overline{V}\}} \\ &= C(V, 0, T_1) \cdot I_{\{V_{T_1} > \overline{V}\}} \cdot I_{\{\tau_{T_1} \notin [0,T_1]\}}. \end{aligned} \tag{6.8}$$

Equation (6.8) differs from equation (6.7) in two important ways:

- First, the maturity of the call (which may be interpreted as an asset claim, since the strike is 0) is T_1 in equation (6.7), and T_2 in equation (6.8). Remembering (cf. equation (4.5), restated for convenience) that

$$\Omega(V, T) = V^{-\beta T},$$

we see immediately that this part of the error is relevant only if the rate β at which the firm generates free cash-flow is strictly positive.

- Second, the last indicator function in equation (6.7) is incorrect: For an asset claim with maturity T_1, possible default at some time $t > T_1$ is irrelevant. Therefore, this term is not contained in equation (6.8).

Given these corrections, we are now in a position to restate Proposition 6.1 in a corrected and extended form, where we include not only calls and heavisides, but also unit down-and-in claims. Moreover, we extend it to allow for an exponential barrier.

Proposition 6.2 (Ericsson/Reneby (1996), corrected and extended). *If default of a corporate security CS as defined in Proposition 4.1 is triggered by at most*

- *a barrier $l_t = l_0 e^{\rho t}$ and*
- *the inability to repay debt at maturity T_2,*

then its value can be expressed as

$$\begin{aligned} CS(\cdot) = {} & a \cdot V \\ & + \sum_{T_1 < T \leq T_2} \sum_i b_{T,i} \cdot C_{lO}(V, F_{T,i}, T, l_0, \rho) \\ & + \sum_{T_1 < T \leq T_2} \sum_i c_{T,i} \cdot H_{lO}(V, F_{T,i}, T, l_0, \rho) \\ & + \sum_{T_1 < T \leq T_2} \sum_i d_{T,i} \cdot G_{lI}(V, T, l_0, \rho) \end{aligned}$$

for $t \leq \tau$ (where $a, b_{(\cdot)}, c_{(\cdot)}, d_{(\cdot)}$ are constants denoting the number of the respective claims), and the time $t = 0$ price of a call option of maturity T_1 and strike K on this corporate security is given by

$$\begin{aligned} C(CS(V, T_2, \cdot), K, T_1) = {} & a \cdot \mathcal{C}_{lO}(V, 0, T_1, l_0, \rho | \{V_{T_1} > \overline{V}\}) \\ & + \sum_{T_1 < T \leq T_2} \sum_i b_{T,i} \cdot \mathcal{C}_{lO}(\cdot | \{V_{T_1} > \overline{V}\}) \\ & + \sum_{T_1 < T \leq T_2} \sum_i c_{T,i} \cdot \mathcal{H}_{lO}(\cdot | \{V_{T_1} > \overline{V}\}) \\ & + \sum_{T_1 < T \leq T_2} \sum_i d_{T,i} \cdot \mathcal{G}_{lI}(\cdot | \{V_{T_1} > \overline{V}\}) \\ & - K \cdot H_{LO}(V, \overline{V}, T_1, l_0, \rho), \end{aligned}$$

where $\overline{V}$ *is defined by*

$$\overline{V}: \ CS_{T_1}(\overline{V}, \cdot) = K,$$

and formulae for $\mathcal{G}_{\mathrm{lI}}$*, the conditional unit down-and-in claim, are given in Lemmata 6.5 and 6.6 below.*

6.2 Reneby (1998)

Reneby (1998) works in an exponential barrier framework. Consequently, he extends the down-and-out probability in Lemma 6.1 to the exponential barrier case:

Lemma 6.3. *Define the event*

$$A_{T_2} = \{V_{T_2} > F, V_{T_1} > \overline{V}, \tau \notin [0, T_2]\}.$$

Then, the associated probabilities under the measures $\mathbb{Q}^m$ *with* $m \in \{V, B, G, G^V\}$ *are given by*

$$\begin{aligned}
\mathbb{Q}^m\{A_{T_2}\} = \; & N\left(d^m_{T_1}\left(\frac{V}{\overline{V}e^{-\rho T_1}}\right), d^m_{T_2}\left(\frac{V}{Fe^{-\rho T_2}}\right), \sqrt{\frac{T_1}{T_2}}\right) \\
& - \left(\frac{L}{V}\right)^{-\frac{2}{\sigma}\cdot\mu^m_X} N\left(d^m_{T_1}\left(\frac{L^2}{V\cdot\overline{V}e^{-\rho T_1}}\right), d^m_{T_2}\left(\frac{L^2}{V\cdot Fe^{-\rho T_2}}\right), \sqrt{\frac{T_1}{T_2}}\right) \\
& - \left(\frac{L}{V}\right)^{-\frac{2}{\sigma}\cdot\mu^m_X} N\left(-d^m_{T_1}\left(\frac{\overline{V}}{Ve^{\rho T_1}}\right), d^m_{T_2}\left(\frac{L^2}{V\cdot Fe^{-\rho T_2}}\right), -\sqrt{\frac{T_1}{T_2}}\right) \\
& + N\left(-d^m_{T_1}\left(\frac{V\cdot\overline{V}}{L^2e^{\rho T_1}}\right), d^m_{T_2}\left(\frac{V}{Fe^{-\rho T_2}}\right), -\sqrt{\frac{T_1}{T_2}}\right)
\end{aligned} \tag{6.9}$$

where d^m_t *and* μ^m_X *are defined as in* (4.15) *and* (4.16).

Using this probability, the prices of conditional down-and-out calls and heavisides in the presence of an exponential barrier are given by

$$\mathcal{C}_{lO}(V, F, T_2, l_0, \rho|\{V_{T_1} > \overline{V}\}) = V\cdot\mathbb{Q}^V(A_{T_2}) - e^{-rT_2}F\cdot\mathbb{Q}^B\{A_{T_2}\}$$

and

$$\mathcal{H}_{lO}(V, F, T_2, l_0, \rho|\{V_{T_1} > \overline{V}\}) = e^{-rT_2}\cdot\mathbb{Q}^B\{A_{T_2}\}.$$

6.3 Extending the Ericsson–Reneby (1996) Results

6.3.1 Lifting Assumptions

In the derivation of their results, Ericsson and Reneby (1996, p. 6) assume that $\overline{V}, F, V > L$ "to make it meaningful". Let us briefly look at these three inequalities separately: The inequality $V > L$ ($V > l_0$ in the exponential barrier case) undoubtedly makes sense: If V were smaller than L (l_0), the company would already be bankrupt at time $t = 0$. The other two restrictions, however, deserve some attention.

Consider first the restriction $F > L$ ($F > l_{T_2}$ in the exponential barrier case). Toft and Prucyk (1997, p. 1153) find within their stationary capital structure setup that the level of the barrier has an important effect on the shape of the equity function. In particular, they show within a simple capital structure consisting only of equity and perpetual coupon debt that, if the barrier is high enough to make the debt essentially risk-free, equity becomes a concave function of asset value. This corresponds, e.g., to the case where $L > F + k$ in a simple capital structure consisting of only equity and finite maturity discount debt within the Ericsson and Reneby (1998) model,[2] where k denotes reorganization costs. Moreover, Leland and Toft (1996) find that for short-term debt, rational equityholders will declare bankruptcy at asset levels that *exceed* debt principal. Therefore, this restriction does not make the analysis "more meaningful". On the contrary, it rules out economically interesting situations.

Let us now look at the restriction $\overline{V} > L$ ($\overline{V} > l_{T_1}$ in the exponential barrier case). If $\overline{V}$ were smaller than L (l_{T_1}), this would simply mean that the option would be exercised for sure at time $t = T_1$ if $\tau \notin [0, T_1]$. Although the mathematics become trivial in this case, it remains unclear why such a situation should not be "meaningful" from an economic point of view.

Extending the Ericsson and Reneby (1996) work requires first an extension of the probabilities given in Lemma 6.1:

Lemma 6.4. *If $F < L$ ($F < l_{T_2}$) in Lemma 6.1, set $F = L$ ($F = l_{T_2}$) when computing $\mathbb{Q}^m\{A_{T_2}\}$. If $\overline{V} < L$ ($\overline{V} < l_{T_1}$) in Lemma 6.1, set $\overline{V} = L$ ($\overline{V} = l_{T_1}$) when computing $\mathbb{Q}^m\{A_{T_2}\}$.*

Then, we have to extend Proposition 6.2:

---

[2] See Section 4.1.

Proposition 6.3. *If* $\overline{V} < L$ *(*$\overline{V} < l_{T_1}$*) in Proposition 6.2, set* $\overline{V} = L$ *(*$\overline{V} = l_{T_1}$*) throughout.*

Proofs. Omitted.

6.3.2 Down-and-Out Underlyings Other than Calls or Heavisides

Throughout this section, $A_{T_1} = \{V_{T_1} > \overline{V}, \tau \notin [0, T_1]\}$.

Definition 6.1. A *unit down-and-in claim* $\mathcal{G}_{LI}(V, T_2, \infty, l_0, \rho | A_{T_1})$, with *infinite maturity* and exponential barrier l_t, starting at time T_2, *conditional on the event* A_{T_1} *and no default in the time interval* (T_1, T_2), is defined via the following payoff function:

$$\Gamma_\tau(\mathcal{G}_{LI}(V, T_2, \infty, l_0, \rho | A_{T_1})) = I_{\{A_{T_1}\}}.$$

Lemma 6.5. *The time* $t = 0$ *price of an infinite maturity conditional unit down-and-in claim as defined above is given by*

$$\mathcal{G}_{LI}(V, T_2, \infty, l_0, \rho | A_{T_1}) = G_{LI}(V, \infty, l_0, \rho) \cdot \mathbb{Q}^G\{A_{T_2}\},$$

where $A_{T_2} = \{A_{T_1}, \tau \notin [0, T_2]\}$*, i.e.,* F *is set equal to* l_{T_2} *when computing* $\mathbb{Q}^G\{A_{T_2}\}$ *from Lemma 6.3.*

Proof. Standard martingale valuation gives

$$\begin{aligned} \mathcal{G}_{LI}(\cdot | A_{T_1}) &= e^{-rT_2} \mathbb{E}^B \left[{}_{T_2}G_{LI}(V, \infty, l_0, \rho) I_{\{A_{T_1}\}} I_{\{\tau \notin [0, T_2]\}} \right] \\ &= e^{-rT_2} \mathbb{E}^B \left[{}_{T_2}G_{LI}(V, \infty, l_0, \rho) I_{\{A_{T_2}\}} \right] \\ &= G_{LI}(V, \infty, l_0, \rho) \cdot \mathbb{Q}^G\{A_{T_2}\}. \end{aligned}$$

□

Definition 6.2. A *unit down-and-in claim* $\mathcal{G}_{LI}(V, T_1, T_2, l_0, \rho | A_{T_1})$, with *finite maturity* and exponential barrier l_t, starting at time T_1, ending at time T_2, *conditional on the event* A_{T_1}, is defined via the following payoff function:

$$\Gamma(\mathcal{G}_{LI}(V, T_1, T_2, l_0, \rho | A_{T_1})) = I_{\{\tau \in (T_1, T_2]\}} I_{\{A_{T_1}\}}.$$

Lemma 6.6 (Reneby 1998). *The time* $t = 0$ *price of a finite maturity conditional unit down-and-in claim as defined above is given by*

$$\mathcal{G}_{LI}(V, T_1, T_2, l_0, \rho | A_{T_1}) = G_{LI}(V, \infty, l_0, \rho) \cdot (\mathbb{Q}^G\{A_{T_1}\} - \mathbb{Q}^G\{A_{T_2}\}),$$

where $A_{T_2} = \{A_{T_1}, \tau \notin [0, T_2]\}$, *i.e.,* F *is set equal to* l_{T_2} *when computing* $\mathbb{Q}^G\{A_{T_2}\}$ *from Lemma 6.3.*

Proof. This claim has already been derived by Reneby (1998, p. 85). The value of this claim must equal a long position in an infinite maturity unit down-and-in claim that starts at time T_1, conditional on A_{T_1}, less another similar claim that starts at time T_2, conditional on A_{T_1} and no default in the time interval (T_1, T_2). □

Definition 6.3. An *exponentially in- or decreasing unit down-and-in claim* $\mathcal{G}^{\nu}_{LI}(V, T_2, \infty, l_0, \rho | A_{T_1})$, with infinite maturity and exponential barrier l_t, starting at time T_2, *conditional on the event* A_{T_1} *and no default in the time interval* (T_1, T_2), is defined via the following payoff function:

$$\Gamma_{\tau}(\mathcal{G}^{\nu}_{LI}(V, T_2, \infty, l_0, \rho | A_{T_1})) = e^{\nu\tau} I_{\{A_{T_1}\}} I_{\{\tau \notin [0, T_2]\}}.$$

Lemma 6.7. *The time* $t = 0$ *price of an infinite maturity conditional exponentially in- or decreasing unit down-and-in claim as defined above is given by*

$$\mathcal{G}^{\nu}_{LI}(V, T_2, \infty, l_0, \rho | A_{T_1}) = G^{\nu}_{LI}(V, \infty, l_0, \rho) \cdot \mathbb{Q}^{G^{\nu}}\{A_{T_2}\},$$

where $A_{T_2} = \{A_{T_1}, \tau \notin [0, T_2]\}$, *i.e.,* F *is set equal to* l_{T_2} *when computing* $\mathbb{Q}^{G^{\nu}}\{A_{T_2}\}$ *from Lemma 6.3. The result holds if* $(\mu^B_X)^2 \geq -2(r - \nu)$.

Proof. Standard martingale valuation gives

$$\begin{aligned}
\mathcal{G}^{\nu}_{LI}(\cdot | A_{T_1}) &= e^{-rT_2} \mathbb{E}^B \left[{}_{T_2}G^{\nu}_{LI}(V, \infty, l_0, \rho) I_{\{A_{T_1}\}} I_{\{\tau \notin [0, T_2]\}} \right] \\
&= e^{-rT_2} \mathbb{E}^B \left[{}_{T_2}G^{\nu}_{LI}(V, \infty, l_0, \rho) I_{\{A_{T_2}\}} \right] \\
&= G^{\nu}_{LI}(V, \infty, l_0, \rho) \cdot \mathbb{Q}^{G^{\nu}}\{A_{T_2}\}.
\end{aligned}$$

□

Remark 6.2. Remember from Remark 4.3 that the definition of this claim is *relative to time* 0 (and, therefore, its value is time-dependent). This time dependence carries over to the finite-maturity version of the claim.

Definition 6.4. An *exponentially in- or decreasing unit down-and-in claim* $\mathcal{G}^{\nu}_{\mathrm{lI}}(V, T_1, T_2, l_0, \rho|A_{T_1})$ with exponential barrier l_t, starting at time T_1, ending at time T_2, *conditional on the event* A_{T_1}, is defined via the following payoff function:

$$\Gamma_{\tau}(\mathcal{G}^{\nu}_{\mathrm{lI}}(V, T_1, T_2, l_0, \rho|A_{T_1})) = e^{\nu\tau} I_{\{A_{T_1}\}} I_{\{\tau \in (T_1, T_2]\}}.$$

Lemma 6.8. *The time* $t = 0$ *price of a finite maturity conditional exponentially in- or decreasing unit down-and-in claim as defined above is given by*

$$\mathcal{G}^{\nu}_{\mathrm{lI}}(V, T_1, T_2, l_0, \rho|A_{T_1}) = G^{\nu}_{\mathrm{lI}}(V, \infty, l_0, \rho) \cdot (\mathbb{Q}^{G^{\nu}}\{A_{T_1}\} - \mathbb{Q}^{G^{\nu}}\{A_{T_2}\}),$$

where $A_{T_2} = \{A_{T_1}, \tau \notin [0, T_2]\}$, *and* $\mathbb{Q}^{G^{\nu}}$ *is the martingale measure with* $G^{\nu}_{\mathrm{lI}}(\cdot, \infty)$ *as numeraire. The associated Girsanov kernel is* $\mu_X^{G^{\nu}} = \mu_X^B - \theta(r - \nu)\sigma$. *The result holds if* $(\mu_X^B)^2 \geq -2(r - \nu)$.

Proof. The value of this claim must equal a long position in an infinite maturity exponentially in- or decreasing unit down-and-in claim that starts at time T_1, conditional on A_{T_1}, less another similar claim that starts at time T_2, conditional on A_{T_2}. □

Remark 6.3. Similar to Lemma 4.16, the sign of ν determines whether the claim is in- or decreasing, and the value of the claim is time-dependent.

Remark 6.4. Similar to the unconditional case (cf. Remark 4.4), the price of the conditional unit down-and-in claim with finite maturity may exist, but not be computable using this formula if the value of its infinite maturity counterpart does not exist.

Definition 6.5. A *down-and-out unit stream* with *infinite maturity*, exponential barrier l_t, starting at time T_2, conditional on the event A_{T_1} and no default in the time interval (T_1, T_2), is defined by the following payoff rate function:

$$\Gamma_t(\mathcal{U}_{\mathrm{lO}}(V, T_2, \infty, l_0, \rho|A_{T_1})) = \Gamma_t(\mathcal{H}_{\mathrm{lO}}(V, 0, t, l_0, \rho|A_{T_1})) \quad t \geq T_2.$$

Lemma 6.9. *The time* $t = 0$ *value of an infinite maturity conditional down-and-out unit stream as defined above is given by*

$$\begin{aligned} \mathcal{U}_{\mathrm{lO}}(V, T_2, \infty, l_0, \rho|A_{T_1}) = \frac{1}{r} \big(& \mathcal{H}_{\mathrm{lO}}(V, 0, T_2, l_0, \rho|A_{T_1}) \\ & - \mathcal{G}_{\mathrm{lI}}(V, T_2, \infty, l_0, \rho|A_{T_1}) \big) . \end{aligned} \tag{6.10}$$

Proof. Standard martingale valuation gives (using equation (4.22))

$$\begin{aligned}
\mathcal{U}_{lO}(\cdot|A_{T_1}) &= e^{-rT_2}\mathbb{E}^B[{}_{T_2}U_{lO}(V,\infty,l_0,\rho)I_{\{A_{T_1}\}}I_{\{\tau\notin[0,T_2]\}}] \\
&= \frac{e^{-rT_2}}{r}\mathbb{E}^B[(1-{}_{T_2}G_{lI}(V,\infty,l_0,\rho))I_{\{A_{T_2}\}}] \\
&= \frac{e^{-rT_2}}{r}(\mathbb{E}^B[I_{\{A_{T_2}\}}]-\mathbb{E}^B[{}_{T_2}G_{lI}(V,\infty,l_0,\rho)]\mathbb{E}^G[I_{\{A_{T_2}\}}]) \\
&= \frac{1}{r}\left(\mathcal{H}_{lO}(V,0,T_2,l_0,\rho|A_{T_1})-G_{lI}(V,\infty,l_0,\rho)\mathbb{Q}^G\{A_{T_2}\}\right) \\
&= \frac{1}{r}\left(\mathcal{H}_{lO}(V,0,T_2,l_0,\rho|A_{T_1})-\mathcal{G}_{lI}(V,T_2,\infty,l_0,\rho|A_{T_1})\right).
\end{aligned}$$

□

Definition 6.6. A *down-and-out unit stream* with *finite maturity*, with exponential barrier l_t, starting at time T_1, ending at time T_2, *conditional on the event* A_{T_1}, is defined by the following payoff rate function:

$$\Gamma_t(\mathcal{U}_{lO}(V,T_1,T_2,l_0,\rho)) = \Gamma_t(\mathcal{H}_{lO}(V,0,t,l_0,\rho|A_{T_1})) \quad t\in[T_1,T_2].$$

Lemma 6.10. *The time* $t=0$ *value of a finite maturity conditional down-and-out unit stream as defined above is given by*

$$\mathcal{U}_{lO}(V,T_1,T_2,l_0,\rho|A_{T_1}) = \mathcal{U}_{lO}(V,T_1,\infty,l_0,\rho|A_{T_1}) - \mathcal{U}_{lO}(V,T_2,\infty,l_0,\rho|A_{T_1}).$$

Proof. The value of this stream must equal a long position in an infinite maturity unit stream that starts at time T_1, conditional on A_{T_1}, less another infinite maturity unit stream starting at time T_2, conditional on A_{T_1}. Note that the conditional down-and-out heaviside in $\mathcal{U}_{lO}(V,T_1,\cdot|A_{T_1})$ collapses to an unconditional down-and-out heaviside. □

Definition 6.7. An *exponentially in- or decreasing down-and-out unit stream* $\mathcal{U}_{lO}^{\nu}(V,T_2,\infty,l_0,\rho|A_{T_1})$, with *infinite maturity* and exponential barrier l_t, starting at time T_2, *conditional on the event* A_{T_1} *and no default in the time interval* $(T_1,T_2]$, is defined via the following payoff rate function:

$$\Gamma_t(\mathcal{U}_{lO}^{\nu}(V,T_2,\infty,l_0,\rho|A_{T_1})) = e^{\nu t}\Gamma_t(\mathcal{H}_{lO}(V,0,t,l_0,\rho|A_{T_1})) \quad t\geq T_2.$$

Lemma 6.11. *The time* $t=0$ *value of an infinite maturity conditional exponentially in- or decreasing down-and-out unit stream as defined above is given by*

$$\begin{aligned}
\mathcal{U}_{lO}^{\nu}(V,T_2,\infty,l_0,\rho|A_{T_1}) = &\frac{1}{r-\nu}\left(e^{\nu T_2}\mathcal{H}_{lO}(V,0,T_2,l_0,\rho|A_{T_1})\right. \\
&\left.-\mathcal{G}_{lI}^{\nu}(V,T_2,\infty,l_0,\rho|A_{T_1})\right).
\end{aligned} \tag{6.11}$$

The result holds if $r > \nu$. *For* $r = \nu$, *the price is found as the limiting case.*

Proof. Standard martingale valuation gives (using equation (4.34) "updated" for the new starting time T_2)

$$\begin{aligned}
\mathcal{U}_{lO}^{\nu}(\cdot|A_{T_1}) &= e^{-rT_2}\mathbb{E}^B\left[{}_{T_2}U_{lO}^{\nu}(V,\infty,l_0,\rho)I_{\{A_{T_1}\}}I_{\{\tau\notin[0,T_2]\}}\right] \\
&= \frac{e^{-rT_2}}{r-\nu}\mathbb{E}^B\left[(e^{\nu T_2} - {}_{T_2}G_{lI}^{\nu}(V,\infty,l_0,\rho))I_{\{A_{T_2}\}}\right] \\
&= \frac{e^{-rT_2}}{r-\nu}\Big((e^{\nu T_2}\mathbb{E}^B[I_{\{A_{T_2}\}}] \\
&\quad -\mathbb{E}^B[{}_{T_2}G_{lI}^{\nu}(V,\infty,l_0,\rho)]\mathbb{E}^{G^\nu}[I_{\{A_{T_2}\}}]\Big) \\
&= \frac{1}{r-\nu}\Big((e^{\nu T_2}\mathcal{H}_{lO}(V,0,T_2,l_0,\rho|A_{T_1}) \\
&\quad - \mathcal{G}_{lI}^{\nu}(V,T_2,\infty,l_0,\rho|A_{T_1})\Big).
\end{aligned}$$

□

Remark 6.5. For $r < \nu$, compare Remark 4.6. The necessary and sufficient condition for the existence of a price in the present case for $r < \nu$ is

$$\mathcal{G}_{lI}^{\nu}(V,T_2,\infty,l_0,\rho|A_{T_1}) > e^{\nu T_2}\mathcal{H}_{lO}(V,0,T_2,l_0,\rho|A_{T_1}).$$

Definition 6.8. An *exponentially in- or decreasing down-and-out unit stream* $\mathcal{U}_{lO}^{\nu}(V,T_2,\infty,l_0,\rho|A_{T_1})$ with *finite maturity* and exponential barrier l_t, starting at time T_1, ending at time T_2, *conditional on the event* A_{T_1}, is defined via the following payoff rate function:

$$\Gamma_t(\mathcal{U}_{lO}^{\nu}(V,T_1,T_2,l_0,\rho|A_{T_1})) = e^{\nu t}\Gamma_t(\mathcal{H}_{lO}(V,0,t,l_0,\rho|A_{T_1})) \quad t \in [T_1,T_2].$$

Lemma 6.12. *The time* $t = 0$ *value of a finite maturity conditional exponentially in- or decreasing down-and-out unit stream as defined above is given by*

$$\begin{aligned}
\mathcal{U}_{lO}^{\nu}(V,T_1,T_2,l_0,\rho|A_{T_1}) = &\,\mathcal{U}_{lO}^{\nu}(V,T_1,\infty,l_0,\rho|A_{T_1}) \\
&- \mathcal{U}_{lO}^{\nu}(V,T_2,\infty,l_0,\rho|A_{T_1}).
\end{aligned} \tag{6.12}$$

This result holds for $r > \nu$. *For* $r = \nu$, *the value of the claim is found as the limiting case.*

Proof. The value of this stream must equal a long position in an infinite maturity exponentially in- or decreasing down-and-out unit stream that starts at time T_1, conditional on A_{T_1}, less another similar claim that starts at time T_2, conditional on A_{T_1}. Note that the conditional down-and-out heaviside in $\mathcal{U}^{\nu}_{lO}(V, T_1, \cdot|A_{T_1})$ collapses to an unconditional down-and-out heaviside. □

Remark 6.6. For $r < \nu$, compare Remark 4.6.

Definition 6.9. A *down-and-out asset stream* with *infinite maturity* and exponential barrier l_t, starting at time T_2, *conditional on the event* A_{T_1} *and no default in the time interval* (T_1, T_2), is defined via the following payoff rate function:

$$\Gamma_t(\mathcal{O}_{lO}(V, T_2, \infty, l_0, \rho|A_{T_1})) = \Gamma_t(\mathcal{C}_{lO}(V, 0, t, l_0, \rho)|A_{T_1}) \quad t \geq T_2. \quad (6.13)$$

Lemma 6.13. *The time* $t = 0$ *value of a conditional, infinite maturity down-and-out asset stream as defined above is (for* $r \neq \rho$*) given by*

$$\begin{aligned}\mathcal{O}_{lO}(V, T_2, \infty, l_0, \rho|A_{T_1}) = \frac{1}{\beta}&(\mathcal{C}_{lO}(V, 0, T_2, l_0, \rho|A_{T_1}) \\ &-l_0 \cdot \mathcal{G}^{\rho}_{lI}(V, T_2, \infty, l_0, \rho|A_{T_1}))\,,\end{aligned} \quad (6.14)$$

where $\mathbb{Q}^{G^\rho}$ *is the martingale measure with* $G^{\rho}_{lI}(\cdot, \infty)$ *as numeraire. The associated Girsanov kernel is* $\mu^{G^\rho}_X = \mu^B_X - \theta(r-\rho)\sigma$. *The result for* $r = \rho$ *is found as the limiting case.*

Proof. Standard martingale valuation gives (using equation (4.24))

$$\begin{aligned}\mathcal{O}_{lO}(\cdot|A_{T_1}) &= e^{-rT_2}\mathbb{E}^B[{}_{T_2}O_{lO}(V, \infty, l_0, \rho)I_{\{A_{T_1}\}}I_{\{\tau\notin[0,T_2]\}}] \\ &= \frac{e^{-rT_2}}{\beta}\mathbb{E}^B[V_{T_2}I_{\{A_{T_2}\}} - l_0 \cdot {}_{T_2}G^{\rho}_{lI}(V, \infty, l_0, \rho)I_{\{A_{T_2}\}}] \\ &= \frac{e^{-rT_2}}{\beta}\Big(\mathbb{E}^B[V_{T_2}]\mathbb{E}^V[I_{\{A_{T_2}\}}] \\ &\quad - l_0\mathbb{E}^B[{}_{T_2}G^{\rho}_{lI}(V, \infty, l_0, \rho)]\mathbb{E}^{G^\rho}[I_{\{A_{T_2}\}}]\Big) \\ &= \frac{1}{\beta}\big(V_0 e^{-\beta T_2}\mathbb{Q}^V\{A_{T_2}\} \\ &\quad -l_0 \cdot G^{\rho}_{lI}(V, \infty, l_0, \rho)\mathbb{Q}^{G^\rho}(A_{T_2})\big) \\ &= \frac{1}{\beta}(\mathcal{C}_{lO}(V, 0, T_2, l_0, \rho|A_{T_1}) - l_0 \cdot \mathcal{G}^{\rho}_{lI}(V, T_2, \infty, l_0, \rho|A_{T_1}))\,.\end{aligned}$$

□

Definition 6.10. A *down-and-out asset stream* with *finite maturity* and exponential barrier l_t, starting at time T_1, ending at time T_2, *conditional on the event* A_{T_1}, is defined via the following payoff rate function:

$$\Gamma_t(\mathcal{O}_{lO}(V, T_1, T_2, l_0, \rho|A_{T_1})) = \Gamma_t(\mathcal{C}_{lO}(V, 0, t, l_0, \rho)|A_{T_1}) \quad t \in [T_1, T_2]. \quad (6.15)$$

Lemma 6.14. *The time* $t = 0$ *value of a conditional, finite maturity down-and-out asset stream as defined above is given by*

$$\mathcal{O}_{lO}(V, T_1, T_2, l_0, \rho) = \mathcal{O}_{lO}(V, T_1, \infty, l_0, \rho|A_{T_1}) - \mathcal{O}_{lO}(V, T_2, \infty, l_0, \rho|A_{T_1}).$$

Proof. Along the lines of the proof of Lemma 6.10. Note that the conditional down-and-out asset claim (call with strike 0) in $O_{lO}(V, T_1, \cdot|A_{T_1})$ collapses to its unconditional counterpart. □

Definition 6.11. A *linearly decreasing unit down-and-in claim* with exponential barrier l_t, starting at time T_1, ending at time T_2, *conditional on the event* A_{T_1}, is defined by the following payoff function:

$$\Gamma_\tau({}^{\text{dec}}_{\text{lin}}\mathcal{G}_{lI}(V, T_1, T_2, l_0, \rho|A_{T_1})) = \frac{T_2 - \tau}{T_2 - T_1} I_{\{A_{T_1}\}} I_{\{\tau \in [T_1, T_2]\}}.$$

Lemma 6.15. *The time* $t = 0$ *value of a conditional linearly decreasing unit down-and-in claim as defined above is given by*

$${}^{\text{dec}}_{\text{lin}}\mathcal{G}_{lI}(V, T_1, T_2, l_0, \rho|A_{T_1}) = G_{lI}(V, \infty, l_0, \rho) \left(\mathbb{Q}^G\{A_{T_1}\} - \frac{\int_{T_1}^{T_2} \mathbb{Q}^G\{A_t\} dt}{T_2 - T_1} \right). \quad (6.16)$$

Proof. From standard martingale valuation and the straightforward extension of Lemma 4.4 to the case of an exponential barrier, we have

$$\begin{aligned} {}^{\text{dec}}_{\text{lin}}\mathcal{G}_{lI}(\cdot) &= \frac{e^{-rT_1}}{T_2 - T_1} \mathbb{E}^B \left[\int_{T_1}^{T_2} {}_{T_1}G_{lI}(V, t, l_0, \rho) I_{\{A_{T_1}\}} dt \right] \\ &= \frac{e^{-rT_1}}{T_2 - T_1} \mathbb{E}^B \left[\int_{T_1}^{T_2} \left({}_{T_1}G_{lI}(V, \infty, l_0, \rho)(I_{\{A_{T_1}\}} - I_{\{A_t\}}) \right) dt \right] \quad (6.17) \\ &= G_{lI}(V, \infty, l_0, \rho) \left(\mathbb{Q}^G\{A_{T_1}\} - \frac{1}{T_2 - T_1} \int_{T_1}^{T_2} \mathbb{Q}^G\{A_t\} dt \right). \end{aligned}$$

□

Remark 6.7. Up to now, we could not find a closed-form solution for the integral of the bivariate normal *distribution* (last term of (6.16)). However, the integral can readily be computed with sufficient precision using standard mathematical software packages.

Definition 6.12. A *down-and-out linearly decreasing unit stream* ${}^{\text{dec}}_{\text{lin}}\mathcal{U}_{lO}(V, T_1, T_2, l_0, \rho|A_{T_1})$ with exponential barrier l_t, starting at time T_1, ending at time T_2, *conditional on the event* A_{T_1}, is defined by the following payoff rate function:

$$\Gamma_t({}^{\text{dec}}_{\text{lin}}\mathcal{U}_{lO}(V, T_1, T_2, l_0, \rho|A_{T_1})) = \frac{T_2 - t}{T_2 - T_1}\Gamma_t(H_{lO}(V, 0, t, l_0, \rho)) \quad t \in [T_1, T_2].$$

Lemma 6.16. *The time* $t = 0$ *value of a conditional linearly decreasing down-and-out unit stream as defined above is given by*

$$\begin{aligned}{}^{\text{dec}}_{\text{lin}}\mathcal{U}_{lO}(V, T_1, T_2, l_0, \rho|A_{T_1}) = \frac{1}{r}\Big(&H_{lO}(V, \overline{V}, T_1, l_0, \rho) \\ &- \frac{1}{(T_2 - T_1)}\mathcal{U}_{lO}(V, T_1, T_2, l_0, \rho|A_{T_1}) \\ &- {}^{\text{dec}}_{\text{lin}}\mathcal{G}_{lI}(V, T_1, T_2, l_0, \rho|A_{T_1})\Big)\,.\end{aligned} \tag{6.18}$$

Proof. From equation (4.39) and standard martingale valuation, we have

$$\begin{aligned}{}^{\text{dec}}_{\text{lin}}\mathcal{U}_{lO}(\cdot|A_{T_1}) &= \frac{e^{-rT_1}}{T_2 - T_1}\mathbb{E}^B\left[\int_{T_1}^{T_2} {}_{T_1}\mathcal{U}_{lO}(V, t, l_0, \rho|A_{T_1})\,dt\right] \\ &= \frac{e^{-rT_1}}{T_2 - T_1}\mathbb{E}^B\left[\int_{T_1}^{T_2}\left({}_{T_1}U_{lO}(V, \infty, l_0, \rho)I_{\{A_{T_1}\}}\right)dt\right. \\ &\quad \left. - \int_{T_1}^{T_2}\left(e^{-r(t-T_1)}{}_tU_{lO}(V, \infty, l_0, \rho)I_{\{A_t\}}\right)dt\right].\end{aligned}$$

Using equation (4.22), we can rewrite that as

$$\begin{aligned}{}^{\text{dec}}_{\text{lin}}\mathcal{U}_{lO}(\cdot|A_{T_1}) &= \frac{e^{-rT_1}}{r(T_2 - T_1)}\mathbb{E}^B\left[\int_{T_1}^{T_2}\Big(\left(1 - {}_{T_1}G_{lI}(V, \infty, l_0, \rho)\right)I_{\{A_{T_1}\}}\right. \\ &\quad \left. - e^{-r(t-T_1)}\left(1 - {}_tG_{lI}(V, \infty, l_0, \rho)\right)I_{\{A_t\}}\Big)dt\right] \\ &= \frac{e^{-rT_1}}{r(T_2 - T_1)}\mathbb{E}^B\left[\int_{T_1}^{T_2}\left(I_{\{A_{T_1}\}} - e^{-r(t-T_1)}I_{\{A_t\}}\right)dt\right. \\ &\quad \left. - \int_{T_1}^{T_2}\left({}_{T_1}G_{lI}(V, \infty, l_0, \rho)(I_{\{A_{T_1}\}} - I_{\{A_t\}})\right)dt\right].\end{aligned}$$

Noting that

$$e^{-rT_1}\mathbb{Q}^B\{A_{T_1}\} = H_{lO}(V,\overline{V},T_1,l_0,\rho),$$

$$\int_{T_1}^{T_2} e^{-r(t-T_1)} I_{\{A_t\}} dt = \int_{T_1}^{T_2} {}_{T_1}\mathcal{H}_{lO}(V,0,t,l_0,\rho|A_{T_1})\, dt$$

$$= {}_{T_1}\mathcal{U}_{lO}(V,T_1,T_2,l_0,\rho|A_{T_1})$$

and remembering from equation (6.17) that

$$\frac{e^{-rT_1}}{(T_2-T_1)}\mathbb{E}^B\left[\int_{T_1}^{T_2}\left({}_{T_1}G_{lI}(V,\infty,l_0,\rho)(I_{\{A_{T_1}\}}-I_{\{A_t\}})\right)dt\right] =$$

$${}^{\text{dec}}_{\text{lin}}\mathcal{G}_{lI}(V,T_1,T_2,l_0,\rho|A_{T_1}),$$

equation (6.18) follows. □

Definition 6.13. A *linearly decreasing, exponentially in- or decreasing unit down-and-in claim* with exponential barrier l_t, starting at time T_1, ending at time T_2, *conditional on the event* A_{T_1}, is defined by the following payoff function:

$$\Gamma_\tau({}^{\text{dec}}_{\text{lin}}\mathcal{G}^\nu_{lI}(V,T_1,T_2,l_0,\rho|A_{T_1})) = e^{\nu\tau}\frac{T_2-\tau}{T_2-T_1} I_{\{A_{T_1}\}} I_{\{\tau\in(T_1,T_2]\}}.$$

Lemma 6.17. *The time* $t=0$ *value of a conditional linearly decreasing, exponentially increasing unit down-and-in claim as defined above is given by*

$${}^{\text{dec}}_{\text{lin}}\mathcal{G}^\nu_{lI}(\cdot|A_{T_1}) = G^\nu_{lI}(V,\infty,l_0,\rho)\left(\mathbb{Q}^{G^\nu}\{A_{T_1}\} - \frac{1}{T_2-T_1}\int_{T_1}^{T_2}\mathbb{Q}^{G^\nu}\{A_t\}\,dt\right). \tag{6.19}$$

This result holds if $(\mu_X^B)^2 \geq -2(r-\nu)$.

Proof. Straightforward extension of the proof of Lemma 6.15. □

Remark 6.8. For the integral in the last term of equation (6.19), compare Remark 6.7.

Remark 6.9. For certain parameter constellations, a price may exist, but not be computable using equation (6.19) (cf. Remark 4.4).

Definition 6.14. A *linearly decreasing, exponentially in- or decreasing down-and-out unit stream* with exponential barrier l_t, starting at time T_1, ending at time T_2, *conditional on the event* A_{T_1}, is defined by the following payoff rate function:

$$\Gamma_t({}^{\text{dec}}_{\text{lin}}\mathcal{U}^{\nu}_{lO}(V, T_1, T_2, l_0, \rho|A_{T_1})) = e^{\nu t}\frac{T_2 - t}{T_2 - T_1}\Gamma_t(H_{lO}(V, 0, t, l_0, \rho)) \quad t \in [T_1, T_2].$$

Lemma 6.18. *The time* $t = 0$ *value of a conditional linearly decreasing, exponentially in- or decreasing down-and-out unit stream as defined above is given by*

$$\begin{aligned}
{}^{\text{dec}}_{\text{lin}}\mathcal{U}^{\nu}_{lO}(V, T_1, T_2, l_0, \rho|A_{T_1}) = \frac{1}{r-\nu}\Big(&e^{\nu T_1}H_{lO}(V, \overline{V}, T_1, l_0, \rho) \\
&- \frac{1}{(T_2 - T_1)}\mathcal{U}^{\nu}_{lO}(V, T_1, T_2, l_0, \rho|A_{T_1}) \qquad (6.20)\\
&- {}^{\text{dec}}_{\text{lin}}\mathcal{G}_{lI}{}^{\nu}(V, T_1, T_2, l_0, \rho|A_{T_1})\Big).
\end{aligned}$$

This result holds for $r > \nu$. *For* $r = \nu$, *the value of the claim is found as the limiting case.*

Proof. From equation (4.44) and standard martingale valuation, we have

$$\begin{aligned}
{}^{\text{dec}}_{\text{lin}}\mathcal{U}^{\nu}_{lO}(\cdot|A_{T_1}) &= \frac{e^{-rT_1}}{T_2 - T_1}\mathbb{E}^B\left[\int_{T_1}^{T_2} {}_{T_1}\mathcal{U}^{\nu}_{lO}(V, t, l_0, \rho|A_{T_1})\,dt\right] \\
&= \frac{e^{-rT_1}}{T_2 - T_1}\mathbb{E}^B\left[\int_{T_1}^{T_2}\left({}_{T_1}U^{\nu}_{lO}(V, \infty, l_0, \rho)I_{\{A_{T_1}\}}\right)dt\right. \\
&\quad \left.- \int_{T_1}^{T_2}\left(e^{-r(t-T_1)}{}_tU^{\nu}_{lO}(V, \infty, l_0, \rho)I_{\{A_t\}}\right)dt\right].
\end{aligned}$$

Using equation (4.34) ("updated" for the new starting time T_1), we can rewrite that as

$$\begin{aligned}
{}^{\text{dec}}_{\text{lin}}\mathcal{U}^{\nu}_{lO}(\cdot|A_{T_1}) &= \frac{e^{-rT_1}}{(r-\nu)(T_2 - T_1)}\mathbb{E}^B\left[\int_{T_1}^{T_2}\left(\left(e^{\nu T_1} - {}_{T_1}G^{\nu}_{lI}(V, \infty, l_0, \rho)\right)I_{\{A_{T_1}\}}\right.\right. \\
&\quad \left.\left.- e^{-r(t-T_1)}\left(e^{\nu t} - {}_tG^{\nu}_{lI}(V, \infty, l_0, \rho)\right)I_{\{A_t\}}\right)dt\right] \\
&= \frac{e^{-rT_1}}{(r-\nu)(T_2 - T_1)}\mathbb{E}^B\left[\int_{T_1}^{T_2}\left(e^{\nu T_1}I_{\{A_{T_1}\}} - e^{-r(t-T_1)+\nu t}I_{\{A_t\}}\right)dt\right. \\
&\quad \left.- \int_{T_1}^{T_2}\left({}_{T_1}G^{\nu}_{lI}(V, \infty, l_0, \rho)(I_{\{A_{T_1}\}} - I_{\{A_t\}})\right)dt\right].
\end{aligned}$$

Noting that

$$e^{-(r-\nu)T_1}\mathbb{Q}^B\{A_{T_1}\} = e^{\nu T_1}H_{lO}(V,\overline{V},T_1,l_0,\rho),$$

$$\int_{T_1}^{T_2} e^{-r(t-T_1)+\nu t} I_{\{A_t\}}\,dt = \int_{T_1}^{T_2} e^{\nu t}\cdot {}_{T_1}\mathcal{H}_{lO}(V,0,t,l_0,\rho|A_{T_1})dt = {}_{T_1}\mathcal{U}^{\nu}_{lO}(V,T_1,T_2,l_0,\rho|A_{T_1}),$$

and remembering from equation (6.17) (extended to the exponentially in- or decreasing claim) that

$$\frac{e^{-rT_1}}{(T_2-T_1)}\mathbb{E}^B\left[\int_{T_1}^{T_2}\left({}_{T_1}G^{\nu}_{lI}(V,\infty,l_0,\rho)(I_{\{A_{T_1}\}}-I_{\{A_t\}})\right)dt\right] = {}^{\text{dec}}_{\text{lin}}\mathcal{G}^{\nu}_{lI}(V,T_1,T_2,l_0,\rho|A_{T_1}),$$

equation (6.20) follows. □

Remark 6.10. For $r < \nu$, compare Remark 4.6.

6.3.3 Put Options on Down-and-Out Underlyings

Suppose, we want to price a put on a down-and-out call, i.e., the right to sell a down-and-out call option (with strike F, expiring at time T_2) at time T_1 for a pre-specified price K. The underlying call will be in-the-money at time T_2 if $V_{T_2} > F$. However, the put will be exercised if $V_{T_1} < \overline{V}$, where $\overline{V}$ is the value of V_{T_1} which makes the value of the underlying call equal to K. In other words, we need the ">/</down-and-out"-probability.

Lemma 6.19 (">/</down-and-out"). *Define the event*

$$A_{T_2} = \{V_{T_2} > F, V_{T_1} < \overline{V}, \tau \notin [0,T_2]\}$$

Then, the associated probabilities under the measures $\mathbb{Q}^m$ *with* $m \in \{B,V,G,G^\nu\}$ *are given by*

$$\mathbb{Q}^m\{A_{T_2}\} = \mathbb{Q}^m\{V_{T_2} > F, \tau \notin [0,T_2]\} - \mathbb{Q}^m\{V_{T_2} > F, V_{T_1} > \overline{V}, \tau \notin [0,T_2]\}, \tag{6.21}$$

where the first probability on the right-hand side in equation (6.21) *is given by Lemma 3.2.*

Proof. By complementarity. □

Other probabilities (e.g., </>/down-and-out) can be derived similarly if needed for different types of contingent claims. Using the probability just derived, the valuation of put options on all types of down-and-out underlyings previously discussed is possible. As an illustration, we give explicitly the price of a put on a down-and-out call:

Lemma 6.20. *The time* $t = 0$ *price of a put option with strike* K *and maturity* T_1 *on a down-and-out call option with strike* F *expiring at time* T_2 *is given by*

$$\begin{aligned} P(C_{lO}(V,F,T_2,l_0,\rho),K,T_1,l_0,\rho) = K \cdot H_{lO}(V,\overline{V},T_1,l_0,\rho) \\ - \mathcal{C}_{lO}(V,F,T_2,l_0,\rho|\{V_{T_1} < \overline{V}\}), \end{aligned} \tag{6.22}$$

where $\overline{V}: C_{lO}(\overline{V},\cdot) = K$*, and the probabilities in the last term are given by Lemma 6.19.*

Proof. Omitted. □

6.3.4 Underlyings of the Up-Barrier Type

If the underlying instrument is not down-and-out, but of some other barrier type (e.g., has an up-barrier), we follow the same approach as in Chapter 3: Derive the corresponding barrier-dependent probability, and price all payoffs using this probability. As an illustration, we will show the derivation of the pricing formula for calls on up-and-out calls and up-and-in calls, respectively, in detail.

We start by deriving the "</</up-and-out-probability".[3]

Lemma 6.21 ("</</up-and-out"). *Let* $V_0 < l_0$*, so that* l_t *represents an up-barrier.* (X_t)*, the default process, is defined as usual:* $X_t = \frac{1}{\sigma}\ln\frac{V_t}{l_t}$*. Define the event* $A_{T_2} = \{V_{T_2} < F, V_{T_1} < \overline{V}, \tau \notin [0,T_2]\}$, ${}_xF = \frac{1}{\sigma}\ln\frac{F}{l_{T_2}}$, ${}_x\overline{V} = \frac{1}{\sigma}\ln\frac{\overline{V}}{l_{T_1}}$, $X_0 = \frac{1}{\sigma}\ln\frac{V_0}{l_0}$*. Then, the probability of the event* A_{T_2} *under*

[3] Although we ultimately need the ">/>/up-and-out-probability", it is more convenient to start with the "</</up-and-out-probability" and derive the former using complementarity relations.

different martingale measures (for $\overline{V} \le l_{T_1}, F \le l_{T_2}$) is given by

$$\begin{aligned}
\mathbb{Q}^m\{A_{T_2}\} = \quad & N\left(\frac{xF - X_0 - \mu_X^m T_2}{\sqrt{T_2}}, \frac{x\overline{V} - X_0 - \mu_X^m T_1}{\sqrt{T_1}}, \sqrt{\frac{T_1}{T_2}}\right) \\
- e^{-2\mu_X^m X_0} \quad & N\left(\frac{xF + X_0 - \mu_X^m T_2}{\sqrt{T_2}}, \frac{x\overline{V} + X_0 - \mu_X^m T_1}{\sqrt{T_1}}, \sqrt{\frac{T_1}{T_2}}\right) \\
- e^{-2\mu_X^m X_0} \quad & N\left(\frac{xF + X_0 - \mu_X^m T_2}{\sqrt{T_2}}, \frac{x\overline{V} - X_0 + \mu_X^m T_1}{\sqrt{T_1}}, -\sqrt{\frac{T_1}{T_2}}\right) \\
+ \quad & N\left(\frac{xF - X_0 - \mu_X^m T_2}{\sqrt{T_2}}, \frac{x\overline{V} + X_0 + \mu_X^m T_1}{\sqrt{T_1}}, -\sqrt{\frac{T_1}{T_2}}\right).
\end{aligned}$$

If $\overline{V} > l_{T_1}$, set $\overline{V} = l_{T_1}$, if $F > l_{T_2}$, set $F = l_{T_2}$ when calculating this probability.

Proof. The proof works along the lines of the proof for Lemma 6.1 (Proposition 3 in Reneby (1998)) and is given to illustrate the general principle. It is a straightforward extension of the derivation of "one time-point probabilities" as shown in the proof of Lemma 3.2.

We are looking for the probability $\mathbb{Q}^m\{V_{T_2} < F, V_{T_1} < \overline{V}, \tau \notin [0, T_2]\}$. From the definition of the Radon–Nikodym derivative, this can be written as

$$\begin{aligned}
\mathbb{Q}^m\{A_{T_2}\} &= \mathbb{E}^X[R^{X\to m} I_{\{X_{T_2} < xF, X_{T_1} < x\overline{V}, \tau \notin [0,T_2]\}}] \\
&= \int_{-\infty}^{xF} \int_{-\infty}^{x\overline{V}} R^{X\to m} \mathbb{Q}^X\{X_{T_2} \in dx_t, X_{T_1} \in dx_t, \tau \notin [0, T_2]\} dx_{T_2} dx_{T_1},
\end{aligned}$$

where x_t denotes a specific realisation of the X_t process.

The probability under $\mathbb{Q}^X$ can then be decomposed by complementarity:

$$\begin{aligned}
\mathbb{Q}^X\{A_{T_2}\} = \mathbb{Q}^X\{&X_{T_2} < xF, X_{T_1} < x\overline{V}\} \\
- \mathbb{Q}^X\{&X_{T_2} < xF, X_{T_1} < x\overline{V}, \tau_{T_1} \in [0, T_1]\} \\
- \mathbb{Q}^X\{&X_{T_2} < xF, X_{T_1} < x\overline{V}, \tau_{T_2} \in [T_1, T_2]\} \\
+ \mathbb{Q}^X\{&X_{T_2} < xF, X_{T_1} < x\overline{V}, \tau_{T_1} \in [0, T_1], \tau_{T_2} \in [T_1, T_2]\},
\end{aligned} \tag{6.23}$$

where τ_{T_1} denotes the first passage time in the interval $[0, T_1]$, and τ_{T_2} denotes the first passage time in the interval $[T_1, T_2]$. The components of equation (6.23) can be found using the reflection principle:

$$\begin{aligned}\mathbb{Q}^X\{A_{T_2}\} = &\ \mathbb{Q}^X\{X_{T_2} < {}_x F, X_{T_1} < {}_x\overline{V}\}\\ &- \mathbb{Q}^X\{X_{T_2} < 2X_0 + {}_xF, X_{T_1} < 2X_0 + {}_x\overline{V}\}\\ &- \mathbb{Q}^X\{-X_{T_2} < 2X_0 + {}_xF, X_{T_1} < {}_x\overline{V}\}\\ &+ \mathbb{Q}^X\{X_{T_2} < {}_xF, -X_{T_1} < 2X_0 + {}_x\overline{V}\}.\end{aligned} \tag{6.24}$$

In terms of bivariate normals, equation (6.24) can be rewritten in integral form as

$$\begin{aligned}\mathbb{Q}^X\{A_{T_2}\} = \int_{-\infty}^{{}_xF}\int_{-\infty}^{{}_x\overline{V}} \Big[& f(0, \sqrt{T_2}, 0, \sqrt{T_1}, \sqrt{T_1/T_2})\\ & - f(-2X_0, \sqrt{T_2}, -2X_0, \sqrt{T_1}, \sqrt{T_1/T_2})\\ & - f(-2X_0, \sqrt{T_2}, 0, \sqrt{T_1}, -\sqrt{T_1/T_2})\\ & - f(0, \sqrt{T_2}, -2X_0, \sqrt{T_1}, -\sqrt{T_1/T_2})\Big]\, dx_{T_2} dx_{T_1}.\end{aligned}$$

Inserting the Radon–Nikodym derivative, changing integration variables to w_t^X, completing the square and integrating gives the probability in the lemma. □

Probabilities for "more complicated" claims can be derived from those given in Lemmata 6.1 and 6.21 by complementarity. For later reference, we state here explicitly the ">/>/up-and-out" and ">/>/up-and-in"-probabilities.

Lemma 6.22 (">/>/up-and-out"). *Let $V_0 < l_0$, so that l_t represents an up-barrier. Define the event*

$$A_{T_2} = \{V_{T_2} > F, V_{T_1} > \overline{V}, \tau \notin [0, T_2]\}$$

Then, the associated probabilities (for $F < l_{T_2}$ and $\overline{V} < l_{T_1}$) under the measures $\mathbb{Q}^m$ are given by

$$\begin{aligned}\mathbb{Q}^m\{A_{T_2}\} = &\ \mathbb{Q}^m\{V_{T_2} < l_{T_2}, V_{T_1} < l_{T_1}, \tau \notin [0, T_2]\}\\ &- \mathbb{Q}^m\{V_{T_2} < l_{T_2}, V_{T_1} < \overline{V}, \tau \notin [0, T_2]\}\\ &- \mathbb{Q}^m\{V_{T_2} < F, V_{T_1} < l_{T_1}, \tau \notin [0, T_2]\}\\ &+ \mathbb{Q}^m\{V_{T_2} < F, V_{T_1} < \overline{V}, \tau \notin [0, T_2]\},\end{aligned} \tag{6.25}$$

where all probabilities are given in Lemma 6.21. If $F > l_{T_2}$ or $\overline{V} > l_{T_1}$, the probability is 0.

Proof. By complementarity. □

Lemma 6.23. *The time* $t = 0$ *price of a call option with strike* K *and maturity* T_1 *on an up-and-out call option with strike* F *and exponential barrier* l_t *expiring at time* T_2 *is given by*

$$C(C_{uO}(V, F, T_2, l_0, \rho), K, T_1, l_0, \rho) = \mathcal{C}_{uO}(V, F, T_2, l_0, \rho | \{V_{T_1} > \overline{V}\}) - K \cdot H_{uO}(V, \overline{V}, T_1, l_0, \rho), \tag{6.26}$$

where $\overline{V} : \ C_{uO}(\overline{V}, \cdot) = K$, *and the probability in the first term on the right-hand side in equation* (6.26) *is given in Lemma 6.22.*

Proof. Omitted. □

Lemma 6.24 (">/>/up-and-in"). *Let* $V_0 < l_0$, *so that* l_t *represents an up-barrier. Define the event*

$$A_{T_2} = \{V_{T_2} > F, V_{T_1} > \overline{V}, \tau \in [0, T_2]\}.$$

Then, the associated probabilities under the measures $\mathbb{Q}^m$ *are given by*

$$\mathbb{Q}^m\{A_{T_2}\} = \mathbb{Q}^m\{V_{T_2} > F, V_{T_1} > \overline{V}\} - \mathbb{Q}^m\{V_{T_2} > F, V_{T_1} > \overline{V}, \tau \notin [0, T_2]\}, \tag{6.27}$$

where the first probability on the right-hand side of equation (6.27) *has been found by Geske (1979) (see equation* (2.27),[4]*) and the second probability is given in Lemma 6.22.*

Proof. By complementarity. □

Lemma 6.25. *The time* $t = 0$ *price of a call option with strike* K *and maturity* T_1 *on an up-and-in call option with strike* F *and exponential barrier* l_t *expiring at time* T_2 *is given by*

$$C(C_{uI}(V, F, T_2, l_0, \rho), K, T_1, l_0, \rho) = \mathcal{C}_{uI}(V, F, T_2, l_0, \rho | \{V_{T_1} > \overline{V}\}) - K \cdot H(V, \overline{V}, T_1, l_0, \rho) \tag{6.28}$$

where $\overline{V} : \ C_{uI}(\overline{V}, \cdot) = K$, *and the probability in the first term on the right-hand side in equation* (6.28) *is given in Lemma 6.24.*

Proof. Omitted. □

[4] Note that this is the limit of the probability given in Lemma 6.1 for $L \to 0$.

Remark 6.11. Note that the heaviside in equation (6.28) is *not* up-and-in. The reason for this is simple: In those cases where the underlying is of the "out"-type, the compound option ceases to exist when the underlying ceases to exist (remember our previous assumption of zero post-default value of corporate securities for the purposes of derivatives valuation). However, in cases where the underlying is of the "in"-type, the compound option (which, itself, is a non-barrier option) may, of course, be exercised regardless of whether its underlying has already come into existence or not.

Chapter 7

Review of Firm Value Based Pricing Models for Equity Derivatives from a Probabilistic Perspective

In this chapter, we review some firm value based option pricing models from a probabilistic perspective. Numerical results will be provided in Chapters 9 and 10. We present the models within the framework of Ericsson and Reneby (1996), using the extensions to this framework derived in Chapter 6. Again, the modularity of this framework not only simplifies the derivation of pricing formulae, but makes economic interpretation of the results much easier compared to formulae found in the literature which have been derived using the PDE approach.

7.1 Option Pricing Extension of Merton (1974): Geske (1979)

Geske (1979) contributes to firm value based option pricing in two important ways: First, using a partial differential equation approach, he derives a closed-form solution for the compound call option. Second, he notes that this approach to option pricing leads to capital structure effects in model prices of options that might potentially explain some of the observed biases of the Black–Scholes model.

To derive the compound call option formula, Geske starts with the Merton (1974) capital structure consisting of equity and one discount bond. His formula can be viewed as a special case of the Ericsson and Reneby (1996) model for this simple capital structure, where in addition $\mathsf{L} = \zeta = 0$, and absolute priority holds in case of bankruptcy (which can only occur at debt maturity).

From equation (6.4) (restated here for convenience), we have

$$C(C_{LO}(V,F,T_2,L),K,T_1) = \mathcal{C}_{LO}(V,F,T_2,L|\{V_{T_1} > \overline{V}\}) \\ - K \cdot H_{LO}(V,\overline{V},T_1,L), \tag{7.1}$$

where A_{T_1} and A_{T_2} collapse to

- $A_{T_1} = \{V_{T_1} > \overline{V}\}$ and
- $A_{T_2} = \{V_{T_2} > F, V_{T_1} > \overline{V}\}$.

The probabilities of these events, given in Lemmata 4.1 and 6.1, collapse to

$$\mathbb{Q}^m\{A_{T_1}\} = N\left(d^m_{T_1}\left(\frac{V_0}{F}\right)\right)$$

and

$$\mathbb{Q}^m\{A_{T_2}\} = N\left(d^m_{T_1}\left(\frac{V}{\overline{V}}\right), d^m_{T_2}\left(\frac{V}{F}\right), \sqrt{\frac{T_1}{T_2}}\right),$$

where d^m_t and μ^m_X are defined as in (4.2) and (4.3). It is easy to verify that equation (7.1) corresponds to equation (2.22) (equation (4) in Geske (1979, p. 68)).

After showing that his compound call formula nests the standard Black–Scholes call price formula as a special case, Geske (1979, p. 73) notes that

> "this result for pricing compound options incorporates the effects of short or long term changes in the firm's capital structure on the value of a call option."

He identifies non-constant equity volatility as the major source for these effects.[1] Geske points out that not only "active" capital structure changes (e.g., issuing a bond) have an impact on option prices, but each move in the stock price will change the debt-to-equity ratio of the firm, thus affecting option prices via a change in stock return volatility.

[1]In effect, not only equity volatility, but the whole distribution of equity returns becomes dependent on the capital structure. This means that, in general, higher moments of this distribution will change as well, also affecting option prices. However, these effects are usually very small compared to the effect induced by a change in volatility.

Geske proceeds by pointing out that the hedge ratio (the "delta") implied by his formula will also reflect this non-constant volatility and shows that it differs from the standard Black–Scholes call delta, which means that a trader hedging according to the Black–Scholes delta will systematically misadjust the hedging portfolio if Geske's model is correct.

He also notes that the direction of the biases of his compound option formula relative to the standard Black–Scholes formula corresponds to empirically observed biases between market prices and Black–Scholes prices: The standard Black–Scholes formula underprices deep-out-of-the-money options and overprices in-the-money options.

7.2 Option Pricing Extension of Leland (1994): Toft and Prucyk (1997)

Toft and Prucyk (1997) extend a version of the Leland (1994) capital structure model to allow for the valuation of standard call and put options on equity within this setting. Their work is based on an extension of the basic Leland model featuring a constant asset payout ratio (see Leland (1994, p. 1241), discussed in Section 5.10). The value of the firm's assets under $\mathbb{P}$ is given by

$$dV_t = (\mu(V_t, t) - \beta)V_t dt + \sigma V_t dW_t.$$

Apart from equity, the only other security is a perpetual bond with continuous coupon payments at rate C per year. Similar to the basic Leland model, coupons are assumed to be tax deductible (the tax rate is denoted by ζ). The barrier triggering bankruptcy is assumed to be a constant L. Toft and Prucyk (1997, p. 1152) point out that one of the main contributions of their paper is the examination of option prices in the presence of an endogenous equity price process.

The value of equity in the extended Leland model is given by (in our notation)

$$E = \Omega_{LO}(V, \infty, L) + \beta O_{LO}(V, \infty, L) - (1 - \zeta)C \cdot U_{LO}(V, \infty, L), \quad (7.2)$$

where, compared to equation (5.17), we impose Leland's assumption regarding the level of the barrier again. As discussed before (see page 104), this effectively assumes that equity will not receive anything in case of default and, thus, restricts permissible barrier levels. Note that the first two terms in equation (7.2) are mutually exclusive: If $\beta > 0$, the first term will be equal to zero; if $\beta = 0$, the second term will be zero.

The price of a call option on equity is then given by

$$\begin{aligned} C(E, K, T_1) = &\ \mathcal{C}_{LO}(V, 0, \infty, L|A_{T_1}) + \beta\mathcal{O}_{LO}(V, \infty, L|A_{T_1}) \\ &- (1-\zeta)C \cdot \mathcal{U}_{LO}(V, \infty, L|A_{T_1}) \\ &- K \cdot H_{LO}(V, \overline{V}, T_1, L), \end{aligned}$$

where A_{T_1} denotes the event $V_{T_1} > \overline{V}, \tau \notin [0, T_1]$, and

$$\overline{V}: \ E_{T_1}(\overline{V}, \cdot) = K.$$

This formula looks quite different from the one derived by Toft and Prucyk (1997) due to different solution techniques. Again, the modular approach in the Ericsson–Reneby framework turns out to be advantageous both in terms of easy derivation and easy interpretation of the formula.[2]

The authors note important deviations when comparing option prices and deltas from their model to those from the Black–Scholes model. First, the structure of the model implies that equity may even be a concave function of asset value for high barrier levels. This implies, e.g., that call deltas do not necessarily converge to zero as $E \to 0$. Computing relative implied Black–Scholes volatilities for options priced correctly within their model, Toft and Prucyk (1997, p. 1163f.) find three important effects:

1. The volatility skew[3] is negatively related to leverage.

2. The volatility skew is more (less) pronounced for options on equity in the presence of a higher (lower) barrier.

3. Differences in volatility skews caused by the barrier level are more pronounced in highly leveraged firms than in firms with low leverage.

[2] Toft and Prucyk (1997, p. 1157) only provide the formula, the derivation is not given (not even in the appendix). They simply mention that "... [the call price formula] can be computed using techniques similar to those used to price barrier options", which might be of little help to readers who do not have much experience with barrier option pricing.

[3] *Volatility skew* refers to the phenomenon that implied volatilities of out-of-the-money calls are lower than those of in-the-money calls. In other words, implied call volatility is a decreasing function of moneyness K/S, with S the underlying and K the strike. This is a well-known "stylized fact" in option markets (see Section 9.1).

One approach to test whether their model potentially explains observed option prices is to conduct an empirical test of hypotheses derived from these numerical findings. The alternative would be to estimate their model for a cross-section of firms. The former approach has two advantages: First, there is no need to estimate the model for each of the firms in the sample, which can be time-consuming and methodologically problematic (cf. the discussion in Ericsson and Reneby (2001, p. 22)). Second, it allows for a more general assessment whether models that are similar in nature to the model discussed here might explain observed market prices. Such models can be expected to show effects that are qualitatively similar, but might not yield exactly the same numbers.

Using a sample of 138 firms with actively traded option contracts, Toft and Prucyk (1997, pp. 1165ff.) run the corresponding regressions and find that the effects predicted from their model are indeed present in their sample. They conclude that capital structure information is an important determinant of observed option prices. This, in turn, provides strong support for further development of firm value based option pricing models.

7.3 Option Pricing Extension of Galai and Schneller (1978): Hanke and Pötzelberger (2002)

Hanke and Pötzelberger (2002) extend the Galai and Schneller (1978) capital structure model for the pricing of options on stocks. Galai and Schneller (1978) analyzed a company whose equity follows a geometric Brownian motion and derived a closed form solution for the prices of warrants issued by this firm. This model can be viewed as a special case of the Ericsson and Reneby (1998) model where the company is financed solely by equity, $L = \zeta = 0$, and, consequently, there is no possibility of default (because there are no "hard payments"). Alternatively, their analysis holds for more complicated capital structures if modelling equity as a geometric Brownian motion is justified.

Suppose, an all-equity financed company with n shares outstanding sells m warrants with maturity T_2 at some time t_0 for an aggregate (fair) price of W_{t_0}. Assuming that the proceeds from selling the warrants are reinvested in projects similar to the firm's existing business, the distribution of (V_t) remains unchanged apart from a level shift reflecting the cash inflow. Thus, we get for the stock price process (S_t) [4]

$$S_t = \begin{cases} \frac{V_t}{n} & \text{for } 0 \leq t < t_0 \\ \left(1 + \frac{mW_{t_0}}{V_{t_0}}\right) \frac{V_t}{n} - \frac{mW_t}{n} & \text{for } t_0 \leq t \leq T_2 \end{cases}. \tag{7.3}$$

From equation (7.3), we see that (as already noted by Galai and Schneller (1978)) the stock of the company after warrants issuance can be regarded as a portfolio, consisting of the stock of an otherwise similar company (without warrants), and a short position in $\frac{m}{m+n}$ calls on this company. Denote the stock price process of this fictitious company with (U_t). Thus,

$$U_t = \left(1 + \frac{mW_{t_0}}{V_{t_0}}\right) \frac{V_t}{n},$$

where W_{t_0} is given by the well-known warrant valuation equation

$$W_{t_0} = \frac{n}{m+n} C_{t_0} \left((1 + \frac{mW_{t_0}}{V_{t_0}}) \frac{V_{t_0}}{n}, x, T_2 \right), \tag{7.4}$$

and we can rewrite equation (7.3) for $t_0 \leq t \leq T_2$ as

$$S_t = U_t - \frac{m}{m+n} C_t(U_t, x, T_2),$$

where x denotes the warrant strike. It is important to note that U_t, although not directly observable, can be calculated from S_t and W_t at any time t.

Since the stock S_t can be regarded as a portfolio consisting of a different stock (U_t) and a short position in $\frac{m}{m+n}$ calls on U_t, a call option on S_t can be viewed as a call on the portfolio. This, in turn, is equivalent to a portfolio of call options on the parts, so they get

$$C_t(S_t, K, T_1) = C_t(U_t, U^*_{T_1}, T_1) - \frac{m}{m+n} CC_t(U_t, K^{cc}, T_1, x, T_2), \tag{7.5}$$

where $U^*_{T_1}$ is chosen such that the call on U is exercised iff the original call on S would be exercised and is given by

$$U^*_{T_1} - \frac{m}{m+n} C_{T_1}(U^*_{T_1}, x, T_2) = K,$$

[4] Since we will only consider options maturing before T_2, we do not concern ourselves with the process after time T_2. See Hanke and Pötzelberger (2002) for further details.

and K^{cc} is chosen such that the compound call will be exercised iff $U_{T_1} > U^*_{T_1}$:

$$K^{cc} = C_{T_1}(U^*_{T_1}, x, T_2).$$

Using the well-known results for the pricing of calls and compound calls in the Black–Scholes setting, this gives us closed-form solutions for options on stocks of companies after warrants issuance.

In this setting, studying effects of warrants issuance on prices of equity options becomes possible. Hanke and Pötzelberger (2002) use data from U.S. companies which issued warrants between 1995 and 2000. They investigate the impact of warrants issuance on several standardized options assumed to be already traded at the time of warrants issuance. Their findings are summarized in Section 10.2.

Chapter 8

Option Pricing Extensions for Several Classical Capital Structure Models

The Geske (1979) model provides an extension of the Merton (1974) capital structure model to allow for the valuation of standard call and put options on stocks. In the same fashion, Toft and Prucyk (1997) provided the corresponding extension to (a variant of) the Leland (1994) capital structure model, and Hanke and Pötzelberger (2002) extended Galai and Schneller (1978).

In this chapter, we will provide option pricing extensions for other classical capital structure models. Their implications for option prices will be examined in chapters 9 and 10. Figure 8.1 shows how the models described in this chapter relate to the existing literature.

8.1 Model 1: Option Pricing Extension of Black and Cox (1976)

The Black and Cox (1976) model (described in Section 5.2) features a capital structure consisting of equity and a discount bond, a constant asset payout rate β, intermediate default triggered by an exponential barrier, but neither bankruptcy costs nor taxes. In case of default, the payoff to equity is 0, which means that debt is not "overprotected": The barrier is never above the present value of debt principal.

Under these conditions, we derived in Section 5.2 the following pricing formula for equity (restated here for convenience):

$$E_0(\cdot) = C_{lO}(V, D, T_2, l_0, \rho) + \beta O_{lO}(V, T_2, l_0, \rho). \tag{8.1}$$

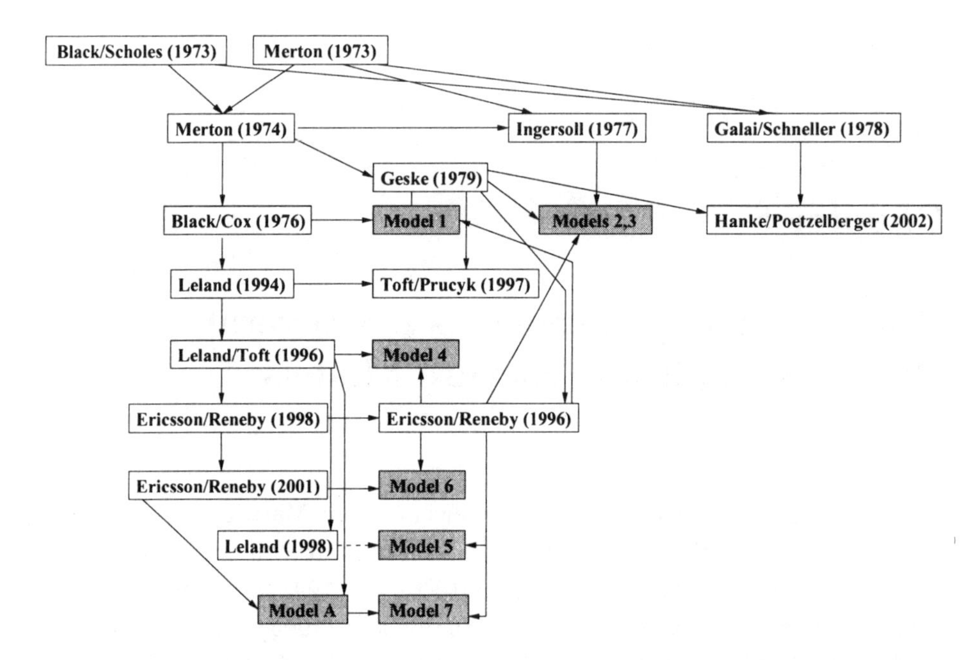

Figure 8.1: Relation of the models described in this book to the existing literature. The dashed arrow from Leland (1998) to Model 5 indicates that Model 5 provides an option pricing extension to a *restricted* version of the Leland (1998) capital structure model.

"Updating" this equation (note that the capital structure is *not* stationary,[1]) we find the value for equity at time T_1 as

$$E_{T_1}(\cdot) = {}_{T_1}C_{lO}(V, D, T_2, l_0, \rho) + \beta \cdot {}_{T_1}O_{lO}(V, T_2, l_0, \rho). \tag{8.2}$$

Proposition 8.1. *The price of a call option on equity with strike* K *and maturity* $T_1 \leq T_2$ *is given by*

$$\begin{aligned} C(E, K, T_1) = {} & \mathcal{C}_{lO}(V, D, T_2, l_0, \rho | \{V_{T_1} > \overline{V}\}) + \beta \mathcal{O}_{lO}(V, T_1, T_2, l_0, \rho | \{V_{T_1} > \overline{V}\}) \\ & - K \cdot H_{lO}(V, \overline{V}, T_1, l_0, \rho), \end{aligned}$$

where $\overline{V}$ *is defined by*

$$E_{T_1}(\overline{V}, \cdot) = K.$$

[1] "Not stationary" means that the value of E_{t_1} and E_{t_2} will be different for $t_1 \neq t_2$ even if $V_{t_1} = V_{t_2}$ (and everything else equal).

Proof. Follows directly from equation (8.2) together with Proposition 6.2. □

8.2 Option Pricing Extensions of Ingersoll (1977a)

8.2.1 Model 2: Convertible Discount Bonds

Ingersoll (1977a) considers capital structures containing callable and non-callable convertible discount bonds. We will focus on non-callable convertibles in this section and defer issues arising from the callability feature to the next section. The model is a non-barrier model extending the Merton (1974) model to *convertible* discount bonds. Thus, default is only possible at maturity, and we do not need barrier-type formulae.

In Section 5.4, we derived the following pricing formula for equity in the presence of a convertible discount bound (equation (5.8), restated here for convenience):

$$E_0(\cdot) = C(V, D, T_2) - \gamma C(V, D/\gamma, T_2), \tag{8.3}$$

where $\gamma = \frac{m}{m+n}$ denoted the dilution factor. "Updating" this equation (note that the capital structure is *not* stationary), we find the value for equity at time T_1 as

$$E_{T_1}(\cdot) = C_{T_1}(V, D, T_2) - \gamma C_{T_1}(V, D/\gamma, T_2), \tag{8.4}$$

Proposition 8.2. *The price of a European call on equity is given by*

$$\begin{aligned} C(E, K, T_1) = &\,\mathcal{C}(V, D, T_2 | \{V_{T_1} > \overline{V}\}) - \gamma \mathcal{C}(V, D/\gamma, T_2 | \{V_{T_1} > \overline{V}\}) \\ &- K \cdot H(V, \overline{V}, T_1), \end{aligned}$$

where $\overline{V}$ *is defined by*

$$E_{T_1}(\overline{V}, \cdot) = K.$$

Proof. Follows directly from equation (8.4) and a simplified (non-barrier) version of Proposition 6.2. □

8.2.2 Model 3: Callable Convertible Discount Bonds

As shown in Section 5.4, under some restrictive assumptions (in particular regarding the growth rate of the call price), a closed-form solution for callable convertible discount bonds (and equity in the presence of callable convertibles) can be derived. We remember that Ingersoll works in a framework without payouts (i.e., $\beta = 0$).

In Section 5.4, we have derived the pricing formula for equity in the presence of a callable convertible discount bond as (cf. equation (5.11), restated here for convenience):

$$E_0(\cdot) = (1-\gamma)\Omega_{uI}(V, T_2, l_0, \rho) + C_{uO}(V, D, T_2, l_0, \rho). \tag{8.5}$$

"Updating" this equation (note that the capital structure is *not* stationary), we find the value for equity at time T_1 as

$$E_{T_1}(\cdot) = (1-\gamma)_{T_1}\Omega_{uI}(V, T_2, l_0, \rho) + {}_{T_1}C_{uO}(V_{T_1}, D, T_2, l_0, \rho). \tag{8.6}$$

Proposition 8.3. *The price of a call option on equity is given by*

$$\begin{aligned} C(E, K, T_1) = (1-\gamma)&\mathcal{C}_{uI}(V, 0, T_2, l_0, \rho|\{V_{T_1} > \overline{V}\}) \\ &+ \mathcal{C}_{uO}(V, D, T_2, l_0, \rho|\{V_{T_1} > \overline{V}\}) - K \cdot H(V_0, \overline{V}, T_1), \end{aligned} \tag{8.7}$$

where $\overline{V}$ *is defined by*

$$E_{T_1}(\overline{V}, \cdot) = K.$$

Proof. Since the "main component" of equity in this case is not down-and-out, but of the up-and-in type, Proposition 6.2 has to be adapted accordingly. The only change necessary is to make the heaviside subtracted in equation (8.7) a *non-barrier* option: Exercise of the call on equity will not be affected by whether conversion has taken place or not (cf. Remark 6.11). □

8.3 Model 4: Option Pricing Extension of Leland and Toft (1996)

The Leland and Toft (1996) model extends the Leland (1994) model for finite maturity debt. In Section 5.11, we have derived the pricing formula for equity (equation (5.23), restated here for convenience) as (remember the restriction $\varphi^D = (1-\varphi^K)$!)

$$\begin{aligned} E_0(\cdot) = V + \zeta C \cdot U_{LO}(V,\infty,L) - \varphi^K L \cdot G_{LI}(V,\infty,L) - \frac{D}{T_2} U_{LO}(V,T_2,L) \\ - C \cdot {}^{\text{dec}}_{\text{lin}} U_{LO}(V,T_2,L) - \varphi^D L \cdot {}^{\text{dec}}_{\text{lin}} G_{LI}(V,T_2,L). \end{aligned} \tag{8.8}$$

"Updating" this equation (note that the capital structure here *is* stationary), we find the value for equity at time T_1 as

$$\begin{aligned} E_{T_1}(\cdot) = V_{T_1} + \zeta C \cdot {}_{T_1}U_{LO}(V,\infty,L) - \varphi^K L \cdot {}_{T_1}G_{LI}(V,\infty,L) \\ - \frac{D}{T_2} \cdot {}_{T_1}U_{LO}(V,T_1+T_2,L) - C \cdot {}^{\text{dec}}_{\text{lin}} U_{LO,T_1}(V,T_1+T_2,L) \\ - \varphi^D L \cdot {}^{\text{dec}}_{\text{lin}} G_{LI,T_1}(V,T_1+T_2,L). \end{aligned} \tag{8.9}$$

Proposition 8.4. *The pricing formula for a call option on equity is given by*

$$\begin{aligned} C(E,K,T_1) = \mathcal{C}_{LO}(V,0,T_1,L|A) + \zeta C \cdot \mathcal{U}_{LO}(V,T_1,\infty,L|A) \\ - \varphi^K L \cdot \mathcal{G}_{LI}(V,T_1,\infty,L|A) - \frac{D}{T_2}\mathcal{U}_{LO}(V,T_1,T_1+T_2,L|A) \\ - C \cdot {}^{\text{dec}}_{\text{lin}}\mathcal{U}_{LO}(V,T_1,T_1+T_2,L|A) \\ - \varphi^D L \cdot {}^{\text{dec}}_{\text{lin}}\mathcal{G}_{LI}(V,T_1,T_1+T_2,L|A) - K \cdot H_{LO}(V,\overline{V},T_1,L), \end{aligned}$$

where A *denotes the event* $\{V_{T_1} > \overline{V}, \tau \notin [0,T_1]\}$, *and* $\overline{V}$ *is defined by*

$$E_{T_1}(\overline{V},\cdot) = K.$$

Proof. Follows directly from equation (8.9) and Proposition 6.2. □

8.4 Model 5: Option Pricing Extension of (a Restricted Version of) Leland (1998)

In the Leland (1998) model, debt is initially issued without a fixed maturity. However, the firm's repayment schedule leads to finite *average* debt maturity. In Section 5.12, we have derived the pricing formula for equity (equation (5.27), restated here for convenience) as

$$\begin{aligned} E_0(\cdot) = V + \zeta C \cdot U_{LO}(V,\infty,L) - \varphi^K L \cdot G_{LI}(V,\infty,L) - k_1 D \\ - k_2 m D \cdot U_{LO}(V,\infty,L) - (C + mD) U_{LO}^{-m}(V,\infty,L) \\ - \varphi^D L \cdot G_{LI}^{-m}(V,\infty,L). \end{aligned} \tag{8.10}$$

"Updating" this equation (note that the capital structure here *is* stationary), we find the value for equity at time T_1 as

$$\begin{aligned} E_{T_1}(\cdot) = V_{T_1} &+ \zeta C \cdot {}_{T_1}U_{LO}(V,\infty,L) - \varphi^K L \cdot {}_{T_1}G_{LI}(V,\infty,L) \\ &- k_2 mD \cdot {}_{T_1}U_{LO}(V,\infty,L) - (C+mD)e^{mT_1} \cdot {}_{T_1}U_{LO}^{-m}(V,\infty,L) \\ &- \varphi^D L \cdot e^{mT_1} \cdot {}_{T_1}G_{LI}^{-m}(V,\infty,L). \end{aligned} \tag{8.11}$$

The factors e^{mT_1} ensure that the respective claims "start at one unit" at time T_1 (remember the time-dependence of these claims discussed, e.g., in Remark 6.2). The cost of initial debt issuance is borne by equityholders at time $t = 0$ and is therefore no longer present in equation (8.11).

Proposition 8.5. *The pricing formula for a call option on equity is given by*

$$\begin{aligned} C(E,K,T_1) = &\, \mathcal{C}_{LO}(V,0,T_1,L|A) + \zeta C \cdot \mathcal{U}_{LO}(V,T_1,\infty,L|A) \\ &- \varphi^K L \cdot \mathcal{G}_{LI}(V,T_1,\infty,L|A) - k_2 mD \cdot \mathcal{U}_{LO}(V,T_1,\infty,L|A) \\ &- e^{mT_1}(C+mD)\mathcal{U}_{LO}^{-m}(V,T_1,\infty,L|A) \\ &- e^{mT_1}\varphi^D L \cdot \mathcal{G}_{LI}^{-m}(V,T_1,\infty,L|A) - K \cdot H_{LO}(V,\overline{V},T_1,L), \end{aligned}$$

where A *denotes the event* $\{V_{T_1} > \overline{V}, \tau \notin [0,T_1]\}$*, and* $\overline{V}$ *is defined by*

$$E_{T_1}(\overline{V},\cdot) = K.$$

Proof. Follows directly from equation (8.11) and Proposition 6.2. □

8.5 Model 6: Option Pricing Extension of Ericsson and Reneby (2001)

The Ericsson and Reneby (2001) model features exponentially increasing debt, with the barrier increasing at the same rate. Here, all debt issues are modelled collectively at an aggregate level. The assumption that single debt issues are too small to affect total debt is convenient for valuing new (small) debt issues and derivatives on these issues, but less appropriate for studying effects of debt issuance on equity options. For completeness, however, we provide the pricing formula for calls on equity in this setting (which is not given in Ericsson and Reneby (2001)).

In Section 4.2, the value of equity was given by the following formula (equation (4.20), restated here for convenience):

$$\begin{aligned} E_0(\cdot) = {} & V - l_0 \cdot G^{\rho}_{lI}(V, \infty, l_0, \rho) - C \cdot U_{lO}(V, \infty, l_0, \rho) \\ & + \zeta C \cdot U^{\rho}_{lO}(V, \infty, l_0, \rho) + \varphi^D \frac{C}{r}(G^{\rho}_{lI}(V, \infty, l_0, \rho) \\ & - G_{lI}(V, \infty, l_0, \rho)) + \varphi^E l_0 G^{\rho}_{lI}(V, \infty, l_0, \rho). \end{aligned} \tag{8.12}$$

"Updating" this equation (note that the capital structure is *not* stationary), we find the value for equity at time T_1 as

$$\begin{aligned} E_{T_1}(\cdot) = {} & V_{T_1} - l_0 \cdot {}_{T_1}G^{\rho}_{lI}(V, \infty, l_0, \rho) - C \cdot {}_{T_1}U_{lO}(V, \infty, l_0, \rho) \\ & + \zeta C \cdot {}_{T_1}U^{\rho}_{lO}(V, \infty, l_0, \rho) + \varphi^D \frac{C}{r}({}_{T_1}G^{\rho}_{lI}(V, \infty, l_0, \rho) \\ & - {}_{T_1}G_{lI}(V, \infty, l_0, \rho)) + \varphi^E l_0 \cdot {}_{T_1}G^{\rho}_{lI}(V, \infty, l_0, \rho). \end{aligned} \tag{8.13}$$

Proposition 8.6. *The pricing formula for a call option on equity is given by*

$$\begin{aligned} C(E, K, T_1) = {} & \mathcal{C}_{LO}(V, 0, T_1, l, \rho|A) - l_0 \cdot \mathcal{G}^{\rho}_{lI}(V, T_1, \infty, l_0, \rho|A) \\ & - C \cdot \mathcal{U}_{lO}(V, T_1, \infty, l_0, \rho|A) + \zeta C \cdot \mathcal{U}^{\rho}_{lO}(V, T_1, \infty, l_0, \rho|A) \\ & + \varphi^D \frac{C}{r}(\mathcal{G}^{\rho}_{lI}(V, T_1, \infty, l_0, \rho|A) - \mathcal{G}_{lI}(V, T_1, \infty, l_0, \rho|A)) \\ & + \varphi^E l_0 \mathcal{G}^{\rho}_{lI}(V, T_1, \infty, l_0, \rho|A) - K \cdot H_{lO}(V, \overline{V}, T_1, l_0, \rho), \end{aligned} \tag{8.14}$$

where A *denotes the event* $\{V_{T_1} > \overline{V}, \tau \notin [0, T_1]\}$, *and* $\overline{V}$ *is defined by*

$$E_{T_1}(\overline{V}, \cdot) = K.$$

Proof. Follows directly from equation (8.13) and Proposition 6.2. □

8.6 Model 7: Option Pricing Extension of Model A

Our "Model A" provides two important extensions to previous models: First, it combines finite debt maturity and exponentially increasing debt in a non-myopic model (i.e., a model which also takes future debt issuance into account). Second, it allows for debt to increase at a rate ν which may be different from the growth rate ρ of the barrier (unlike the Ericsson and Reneby (2001) model, where the two are assumed to be equal).

In Section 5.13, we derived the value of equity as

$$\begin{aligned} E_0(\cdot) = {} & V + \zeta C \cdot U^{\nu}_{lO}(V, \infty, l_0, \rho) - \varphi^K L \cdot G^{\rho}_{lI}(V, \infty, l_0, \rho) \\ & - \frac{D}{T_2} U^{\nu}_{lO}(V, T_2, l_0, \rho) - C \cdot {}^{dec}_{lin}U^{\nu}_{lO}(V, T_2, l_0, \rho) \\ & - \varphi^D l_0 \cdot {}^{dec}_{lin}G^{\nu}_{lI}(V, T_2, l_0, \rho). \end{aligned} \tag{8.15}$$

"Updating" this equation (note that debt is increasing over time), we find the value for equity at time T_1 as

$$\begin{aligned} E_{T_1}(\cdot) = {} & V_{T_1} + \zeta C \cdot {}_{T_1}U^{\nu}_{lO}(V, \infty, l_0, \rho) \\ & - \varphi^K l_{T_1} \cdot {}_{T_1}G^{\rho}_{lI}(V, \infty, l_0, \rho) - \frac{D}{T_2} \cdot {}_{T_1}U^{\nu}_{lO}(V_{T_1}, T_1 + T_2, l_0, \rho) \\ & - C \cdot {}^{dec}_{lin}U^{\nu}_{lO,T_1}(V, T_1 + T_2, l_0, \rho) \\ & - e^{-\nu T_1} \varphi^D l_{T_1} \cdot {}^{dec}_{lin}G^{\nu}_{lI,T_1}(V, T_1 + T_2, l_0, \rho). \end{aligned} \tag{8.16}$$

Proposition 8.7. *The pricing formula for a call option on equity is given by*

$$\begin{aligned} C(E, K, T_1) = {} & \mathcal{C}_{LO}(V, 0, T_1, l_0, \rho|A) + \zeta C \cdot \mathcal{U}^{\nu}_{lO}(V, T_1, \infty, l_0, \rho|A) \\ & - \varphi^K L \cdot \mathcal{G}^{\rho}_{lI}(V, T_1, \infty, l_0, \rho|A) - \frac{D}{T_2} \mathcal{U}^{\nu}_{lO}(V, T_1, T_1 + T_2, l_0, \rho|A) \\ & - C \cdot {}^{dec}_{lin}\mathcal{U}^{\nu}_{lO}(V, T_1, T_1 + T_2, l_0, \rho|A) \\ & - e^{-\nu T_1} \varphi^D L \cdot {}^{dec}_{lin}\mathcal{G}^{\nu}_{lI}(V, T_1, T_1 + T_2, l_0, \rho|A) \\ & - K \cdot H_{lO}(V, \overline{V}, T_1, l_0, \rho), \end{aligned} \tag{8.17}$$

where A *denotes the event* $\{V_{T_1} > \overline{V}, \tau \notin [0, T_1]\}$, *and* $\overline{V}$ *is defined by*

$$E_{T_1}(\overline{V}, \cdot) = K.$$

Proof. Follows directly from equation (8.16) and Proposition 6.2. □

Chapter 9

Capital Structure Effects in Option Prices – The Static Case

In this chapter, we investigate the properties of option prices within some of the firm value based pricing models described in Chapters 7 and 8. First steps into this direction have been made by Ericsson and Reneby (1996, pp. 13ff.). Our approach will be to describe the biases of prices calculated within these models relative to prices calculated from the Black–Scholes formula. There are two main reasons for taking this approach:

- A number of biases between market prices of options and Black–Scholes prices are well-documented in the literature. If the firm value based models do a better job at "explaining" market prices compared to the Black–Scholes formula, pricing biases between these models and the Black–Scholes model should resemble those between market prices and the Black–Scholes formula.

- Practitioners all over the world use the Black–Scholes formula every day. Of course, they are well aware of its biases and adjust the inputs to the formula to account for these biases. Quite often, e.g., prices are quoted in terms of implied Black–Scholes volatilities (i.e., the volatilities which, when used as an input to the Black–Scholes formula, make Black–Scholes prices equal market prices), and traders regularly use what is known as the "implied volatility surface" (see below) for pricing purposes. Thus, a practical understanding of the option pricing implications of firm value based models should be made easier by an investigation of their biases relative to the Black–Scholes model (compared to the alternative of examining "raw" prices).

9.1 Pricing Biases of the Black–Scholes Model – "Stylized Facts"

In the empirical option pricing literature, two main pricing biases between market prices of options and Black–Scholes prices have been documented (see, e.g., Rubinstein (1985) or Dumas, Fleming, and Whaley (1998)):

1. (Asymmetric) volatility smile
2. Term structure of volatilities

9.1.1 The Volatility Smile

The term "volatility smile" refers to the phenomenon that Black–Scholes option prices deviate systematically from market prices, depending on the moneyness ratio (K/U, where U denotes the value of the underlying). For some underlyings (e.g., foreign currency), the Black–Scholes formula underprices at-the-money options ($K/U \approx 1$) relative to in-the-money options (for calls: $K/U < 1$) and out-of-the-money options (for calls: $K/U > 1$), which leads to a pattern resembling a human smile when plotting implied volatilities against moneyness (see Figure 9.1).

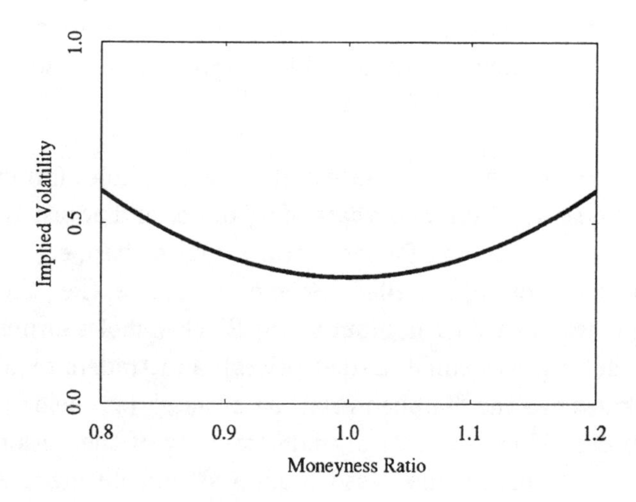

Figure 9.1: Typical (symmetric) volatility smile for options on foreign currency

For equity, however, the pattern looks markedly different: The Black–Scholes formula overprices in-the-money options relative to at-the-money options, which, in turn, are overpriced relative to out-of-the-money options. Graphically, this pattern is described as a "sneer" (or "asymmetric smile", see Figure 9.2).

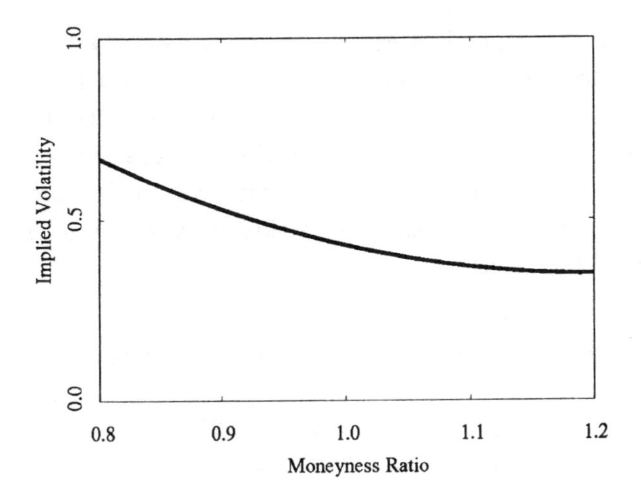

Figure 9.2: Typical (asymmetric) volatility smile for options on stocks

9.1.2 The Term Structure of Volatilities

"Term structure of volatilies" refers to the phenomenon that implied Black–Scholes volatilities are not constant across option maturities (for options with the same strike price). A changing equity volatility is usually thought to be the most important cause for this phenomenon. An upward-sloping (downward-sloping) term structure is then interpreted as the market predicting an increase (decrease) in volatility. Although this phenomenon is well-known, there does not seem to exist much empirical work related to it.

When implied volatilities are plotted against the two dimensions *moneyness ratio* and *option maturity*, the resulting graph is usually referred to as an *implied volatility surface*.

9.1.3 The Debt-Maturity Term Structure of Volatilities

The dependence of option prices on debt maturity has been investigated very briefly by Ericsson (1997, p. 17f.). To the best of our knowledge, empirical results relating (equity) option pricing biases to debt maturity do not exist. This is hardly surprising given that debt maturity is not an input to the Black–Scholes formula (as opposed to strike and option maturity).

However, possible dependencies between debt maturity and equity option prices can be investigated in a firm value-based pricing framework. Therefore, we include this third bias type in our numerical analyses, although the motivation for this stems from the increased flexibility offered by those models rather than empirical observations. In the absence of any existing terminology, we propose to call this relation the *debt-maturity term structure of volatilities.*

9.2 Pure Debt-Equity Capital Structures

In this section, we investigate numerically pricing biases between the Black–Scholes formula and a subset of the models described in previous sections, namely those where the capital structure consists only of debt and equity. The other models will be dealt with in Section 9.3. For the pure debt-equity models, there exists a "hierarchy" in the sense that Model A nests most of the other models. With each model, we follow a standardized approach: We assume that the model under investigation represents the "true world", and compute option prices using the formulae derived in Chapters 7 and 8 (which, given our assumption, are correct). Then, we use these prices to compute implied Black–Scholes volatilies, i.e., those volatility inputs which would yield the correct option prices when using the Black–Scholes formula.

To develop an understanding for the qualitative effects and to avoid an excessive number of tables, we use figures to illustrate our findings for the economically less rich "classical models". However, we will provide exact numerical results for Model A/Model 7.

9.2.1 Merton (1974)/Geske (1979)

For the Merton (1974)/Geske (1979) model, we start with the following parameters ("base case"): $V = 100$, $r = 0.08$, $\sigma = 0.2$, $T_2 = 10$, and a leverage of 40% (where leverage is defined as $D(\cdot)/(D(\cdot)+E(\cdot))$, and $D(\cdot)$ and $E(\cdot)$ denote the market prices of debt and equity, respectively). The corresponding debt principal is $D = 92.29$, yielding an equity value of $E(\cdot) = 60$.

For an at-the-money call ($K = E(\cdot)$) with expiration time $T_1 = 0.25$, we get a price of 4.41. This gives us an implied Black–Scholes volatility of 0.32.

9.2.1.1 Volatility Smile

Repeating these calculations for strikes between 80% and 120% of equity, we get the implied volatility smile in Figure 9.3. We see that the shape of

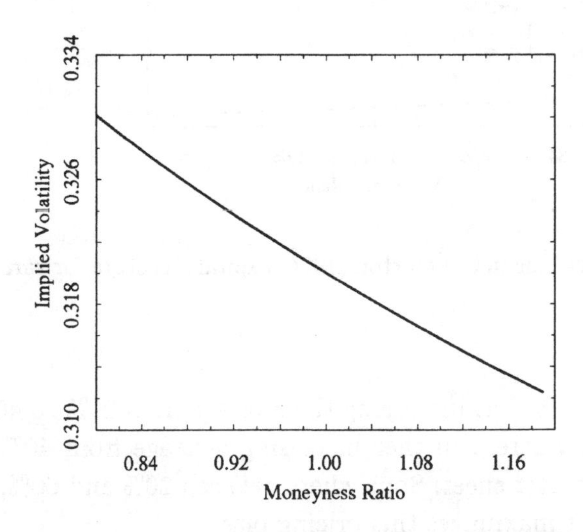

Figure 9.3: Moneyness bias in the Merton (1974) capital structure (base case)

the bias corresponds to the empirical findings discussed in the previous section: Assuming that the Geske (1979) option prices are correct, the implied Black–Scholes volatilities show the typical "sneer" pattern.

Next, we want to examine the dependence between this pattern and some of the input parameters. We start by varying the leverage, considering levels of 20% and 60% in addition to the base case of 40%. Debt maturity is kept constant at $T_2 = 10$. 20% leverage corresponds to a debt principal of $D = 44.62$, whereas 60% leverage corresponds to a debt principal of

D = 152.89. Figure 9.4 shows that – not surprisingly – higher leverage leads to higher implied option volatilities (since a higher leverage increases equity volatility). Standardizing by the at-the-money implied volatilities,

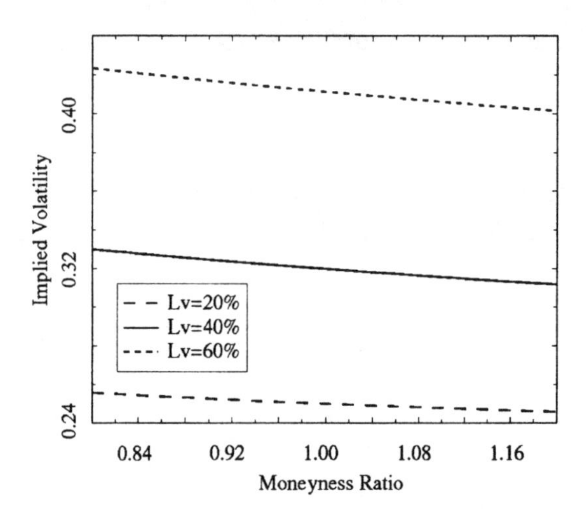

Figure 9.4: Moneyness bias in the Merton (1974) capital structure for various levels of leverage (Lv)

we see from Figure 9.5 that increasing the leverage from 20% to 40% leads to a steeper sneer, whereas further increasing leverage from 40% to 60% leads to a slightly flatter sneer: Somewhere between 20% and 60%, there is a leverage level that maximizes this pricing bias.

We continue by varying the option's maturity. Figure 9.6 shows that changes in option maturity lead to an almost exact parallel shift in the smile, whereby longer option maturities lead to slightly higher implied volatilities. The shape of the sneer is hardly affected by variations in the option maturity.

Next, we want to examine the dependence of the sneer on debt maturity (keeping leverage constant at 40%). The corresponding values for debt principal are D = 59.99 (for a debt maturity of $T_2 = 5$) and D = 228.42 (for a debt maturity of $T_2 = 20$). Figure 9.7 shows that, not surprisingly, longer debt maturities lead to lower implied volatilities. However, not only the level of implied volatilities is affected by debt maturity, but also the

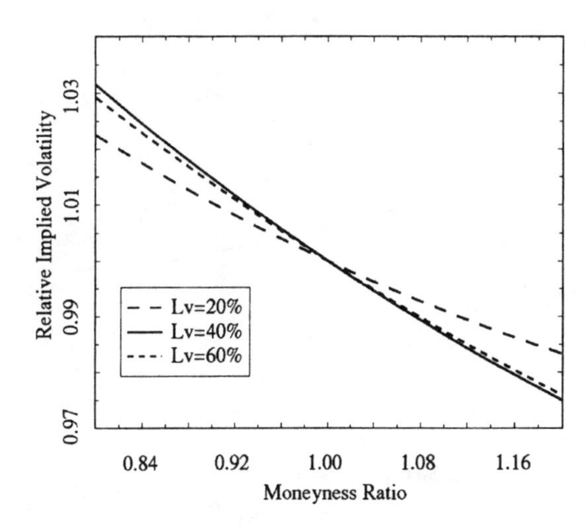

Figure 9.5: Relative implied volatilies in the Merton (1974) capital structure for various levels of leverage (Lv)

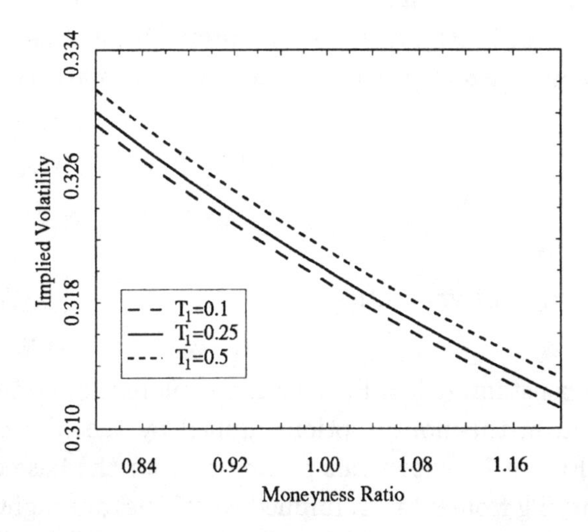

Figure 9.6: Moneyness bias in the Merton (1974) capital structure for various option maturities

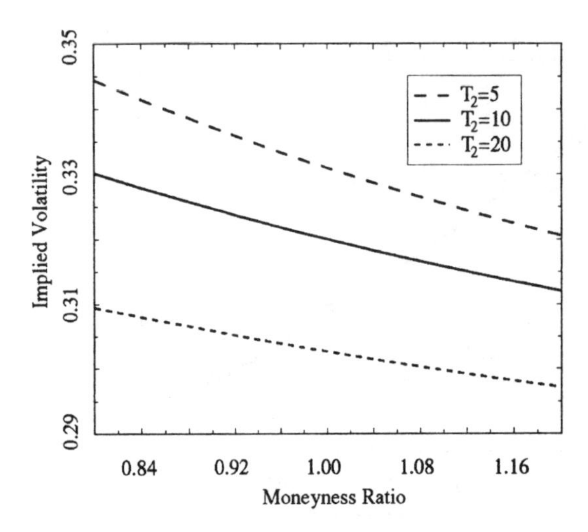

Figure 9.7: Moneyness bias in the Merton (1974) capital structure for various debt maturities

shape of the volatility sneer. This can best be seen by standardizing the implied volatility curves in Figure 9.7 by the respective at-the-money implied volatilities. Figure 9.8 confirms that shorter debt maturities lead to steeper sneers.

9.2.1.2 Term Structure of Volatilities

Keeping the strike constant at $K = E(\cdot)$, we can explore the term structure of volatilities for an at-the-money option implied by the Merton (1974) capital structure. Figure 9.9 shows this pricing bias for the base case with option maturity varying from 0.1 to 1. Implied volatilities are slightly higher for options with longer maturities, although the effect is *very* small. The shape of this pricing bias is affected by variations in the input parameters, but the overall effect remains negligible for practical purposes, and we omit a detailed analysis for this reason.

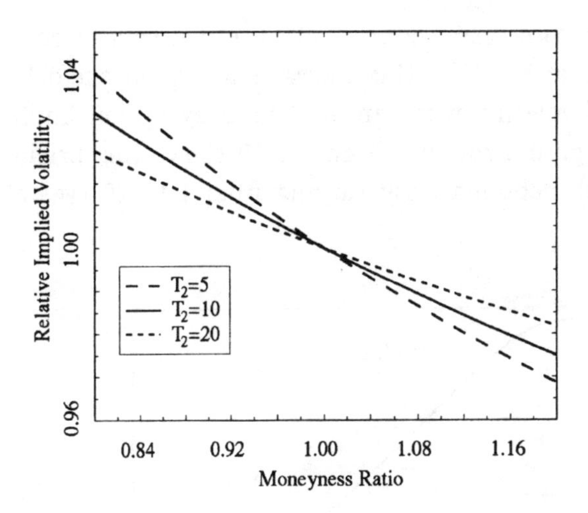

Figure 9.8: Relative implied volatilies in the Merton (1974) capital structure for various debt maturities

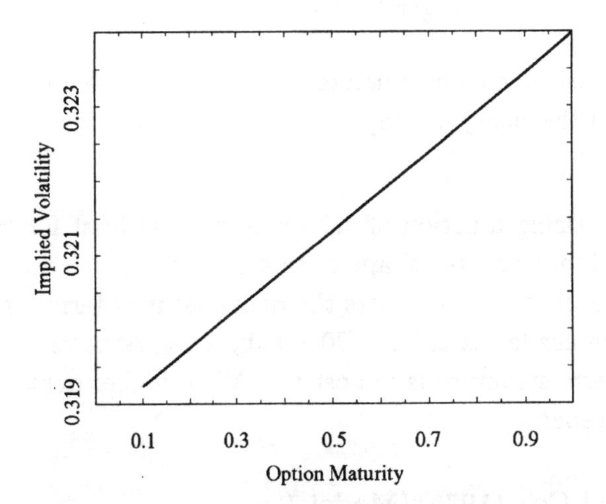

Figure 9.9: Term structure of volatilities in the Merton (1974) capital structure (at-the-money option)

9.2.1.3 Debt-Maturity Term Structure of Volatilities

Starting from our base case, we now vary the maturity of debt. Keeping the strike constant at $K = E(\cdot)$, this allows us to explore the debt-maturity term structure of volatilities for an at-the-money option implied by the Merton (1974) capital structure. Figure 9.10 shows this pricing bias for the base case with debt maturity varying from 1 to 20 (years). Implied

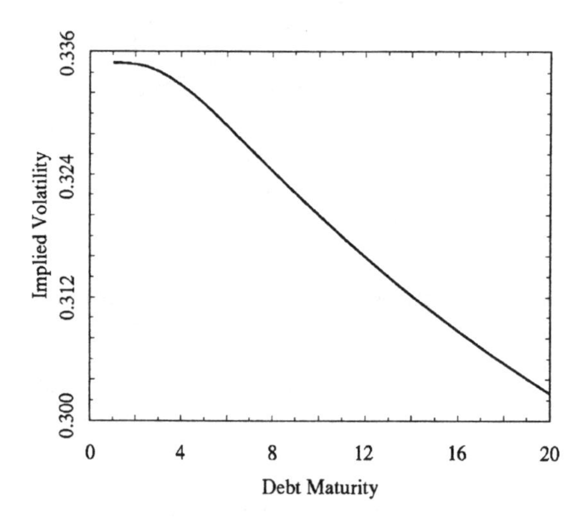

Figure 9.10: Debt-maturity term structure of volatilities in the Merton (1974) capital structure (at-the-money option)

volatility is a decreasing function of debt maturity (at least for "realistic" debt maturities). Moreover, the shape of this pricing bias is affected by the level of leverage. Figure 9.11 contrasts the debt-maturity term structure of volatilities for leverage levels of $Lv = 20, 40$ and 60%, respectively. For low debt levels, the term structure is almost flat. With higher debt levels, the curve becomes steeper.

9.2.2 Black and Cox (1976)/Model 1

As discussed in Section 5.2, the Black and Cox (1976) model extends Merton's framework in two ways: First, it allows for payouts by the firm at a constant rate β. Second, it provides for intermediate default, triggered

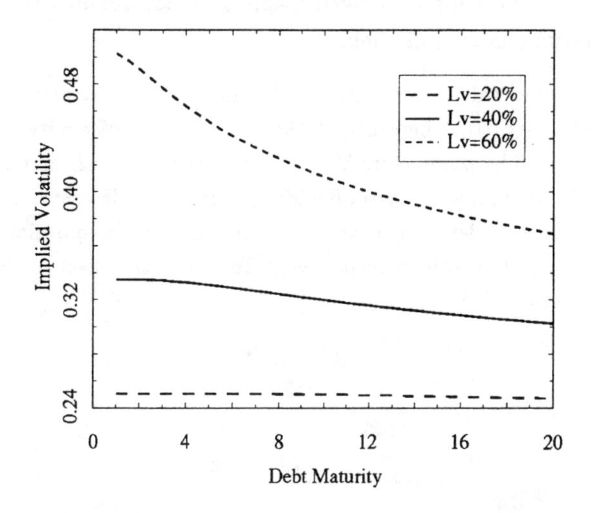

Figure 9.11: Debt-maturity term structure of volatilities in the Merton (1974) capital structure for various levels of leverage (at-the-money option)

by the first passage of the firm value to an exponential barrier l_t. In other words, the Merton (1974) framework is a special case of the Black and Cox (1976) model with $\beta = l_t = 0$.

We will investigate the effects of $\beta \neq 0$ and $l_t \neq 0$ separately, and start with a non-zero payout.

9.2.2.1 Positive Asset Payouts, No Barrier

We use the same base case parameters as in the previous section ($V = 100$, $Lv = 40\%$, $r = 0.08$, $\sigma = 0.2$, $T_1 = 0.25$ and $T_2 = 10$). The dividend yield q used in the Black–Scholes formula for computing the implied volatility is computed from (cf. Toft and Prucyk (1997, p. 1162))

$$F = E(\cdot)^{(r-q)T_1},$$

where F denotes the forward price of equity (i.e., the value of a call option with strike 0). The asset payout rate β is chosen such that the (annualized) expected dividend yield for equity over the option's lifetime is 3%. For our base case, this implies a debt principal of $D = 95.04$ and a payout rate of $\beta = 1.79\%$. The corresponding equity value is, of course, 60 (since total

Table 9.1: Parameter Values for Leveraged Capital Structures in the Black–Cox (1976) Model for Various Dividend Yields q

This table shows parameter values for 40 percent leveraged capital structures. Leverage is defined as $D(\cdot)/(D(\cdot)+E(\cdot))$, where $E(\cdot)$ is the market value of equity, and $D(\cdot)$ is the market value of debt. The parameters $V = 100$, $r = 0.08$, $\sigma = 0.2$, and $T_2 = 10$ are identical for all capital structures. The asset liquidation rate, β, is determined so that the dividend payable to equity holders during the next three months is equivalent to a 0, 3, and 6 percent dividend yield on equity, respectively. To isolate the effect of the dividend yield q, the barrier is kept at 0.

	q (div. yield)		
	0	3%	6%
D	92.29	95.04	99.92
β(%)	0	1.79	3.56

firm value is just the sum of the market values of equity and debt in the absence of taxes). For an at-the-money call, we get a price of 8.65, which gives an implied volatility of 31.67%. Table 9.1 summarizes the parameter values for 40% leveraged capital structures with dividend yields of 0, 3, and 6%.

First, we examine the impact of a non-zero dividend yield on the volatility smile. Figure 9.12 shows that a higher dividend yield leads to a decrease in implied volatility. Moreover, the sneer becomes flatter for higher values of q, which can be seen more easily from Figure 9.13. Here, the values from Figure 9.12 have been standardized by the respective at-the-money implied volatilities.

Next, we explore the impact of positive payouts on the term structure of volatilities. To facilitate comparison with the Merton (1974) model, we focus again on the at-the-money option and include the case of $q = \beta = 0$ (which corresponds to Figure 9.9). Figure 9.14 shows that a higher dividend yield leads to lower implied volatilities. However, the shape of the term structure of volatilities is hardly affected and upward-sloping in all cases.

Finally, we look at the impact of q on the debt-maturity term structure of volatilities. Extending the results shown in Figure 9.10 in the now already familiar way, we see from Figure 9.15 that higher asset payout ratios lead to slightly higher implied volatilities for extremely short-term debt

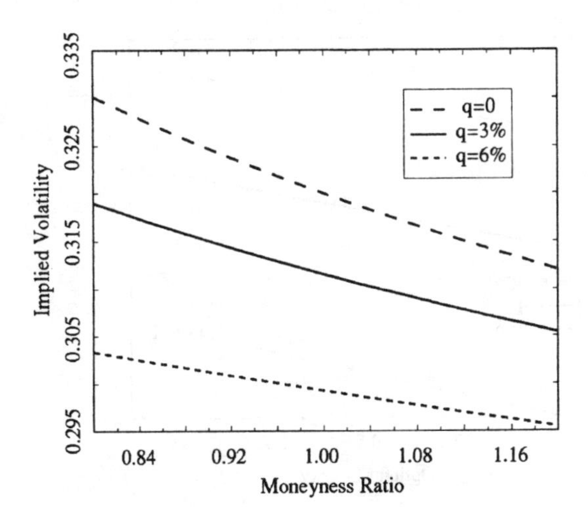

Figure 9.12: Moneyness bias in the Black–Cox (1976) capital structure for various levels of the dividend yield q

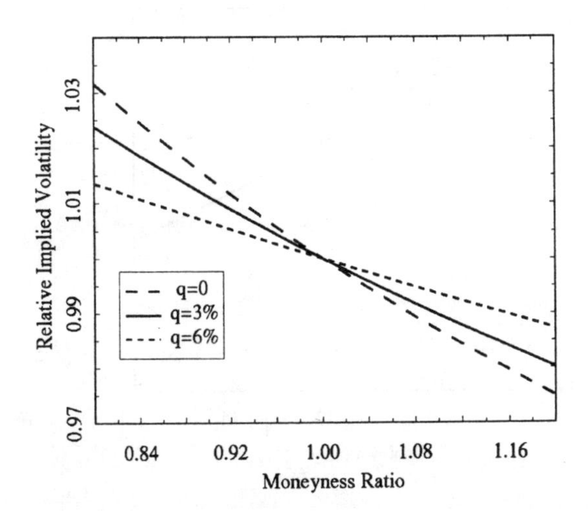

Figure 9.13: Relative implied volatilies in the Black–Cox (1976) capital structure for various levels of the dividend yield q

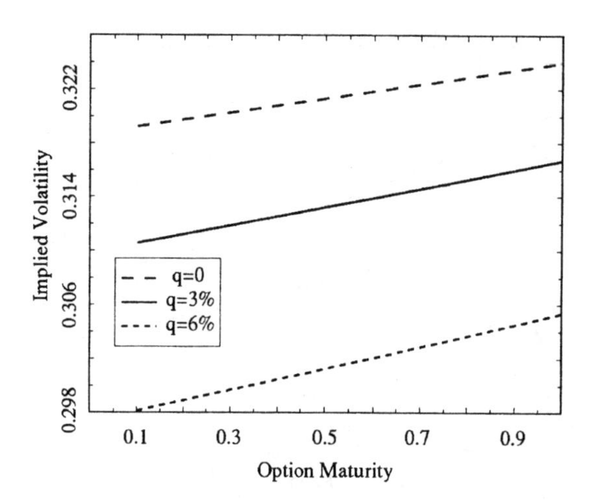

Figure 9.14: Term structure of volatilities in the Black–Cox (1976) capital structure for various levels of the dividend yield q (at-the-money option)

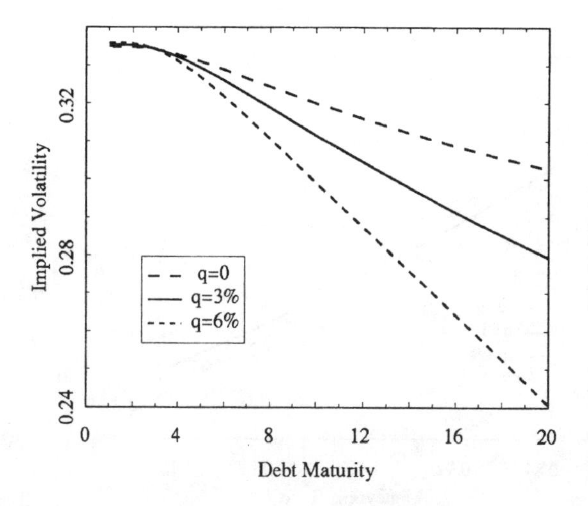

Figure 9.15: Debt-maturity term structure of volatilities in the Black–Cox (1976) capital structure for unprotected debt for various levels of the dividend yield q (at-the-money option)

(with a maturity of less than about 2 years), and markedly lower implied volatilities for medium- and long-term debt. Moreover, the debt-maturity term structure becomes steeper for higher payout ratios.

9.2.2.2 Positive and Exponentially Increasing Barrier, No Asset Payouts

In this subsection, we isolate the effect on Black–Scholes pricing biases of extending the Merton–Geske framework for an exponential barrier. Remember from Section 5.2 that, to avoid "debt overprotection", Black and Cox (1976) assume that (in our notation) $\mathfrak{l}_0 e^{\rho t} \leq D e^{-r(T_2-t)}$ holds for all $0 \leq t \leq T_2$. We use the same base case parameters as in the previous subsection, with the exception of $\beta = 0$. In addition, we set the base case value of the barrier to $\mathfrak{l}_0 = 0.6 \cdot D e^{-rT_2}$ and $\rho = r$. This ensures that the recovery ratio of debt is 60% of the present value of the promised final payment in case of default at $\tau < T_2$ (cf. Black and Cox (1976, p. 357)).

For our base case, this implies a debt principal of $D = 92.25$, and a barrier value at time 0 of $\mathfrak{l}_0 = 24.87$. The corresponding equity value is, of course, 60 (since total firm value is just the sum of the market values of equity and debt in the absence of taxes). For the call, we get a price of 4.41, which gives an implied volatility of 32.03%. Table 9.2 summarizes the parameter values for 40% leveraged capital structures with debt protection levels of 60, 80 and 100%.

First, we look at the effect of the barrier level on the smile. From Figure 9.16, we see that a higher barrier (corresponding to higher debt recovery ratios) leads to higher implied volatilities. The impression that a higher barrier also leads to a steeper sneer is confirmed by Figure 9.17, where the implied volatility curves have been standardized by their respective at-the-money values. The effect on the term structure of volatilities is confined to a small level shift, but no change in shape relative to the Merton–Geske model (result not shown).

Turning to the debt maturity term structure of volatilities, Figure 9.18 shows an interesting effect: Whereas for low debt protection (60% of present value of promised payout), the debt maturity term structure looks similar to that in case of no debt protection (cf. Figure 9.15), the curve becomes flatter at the long end for higher barrier values. For very high values of the barrier (around 80% for our base case parameters), the curve even starts to increase at high debt maturities.

Table 9.2: Parameter Values for Leveraged Capital Structures in the Black–Cox (1976) Model for Various Debt Protection Levels

This table shows parameter values for 40 percent leveraged capital structures. Leverage is defined as $D(\cdot)/(D(\cdot) + E(\cdot))$, where $E(\cdot)$ is the market value of equity, and $D(\cdot)$ is the market value of debt. The parameters $V = 100$, $r = 0.08$, $\sigma = 0.2$, and $T_2 = 10$ are identical for all capital structures. The barrier is chosen such that debt protection (i.e., the recovery ratio for debtholders in case of default) is constant at 60, 80 and 100%, respectively, of the time value of debt. This implies that the barrier grows at rate r. To isolate the effect of the barrier, the dividend yield q (or, viewed alternatively, the asset payout ratio β) is kept at 0.

	Debt protection		
	60%	80%	100%
D	92.25	91.65	89.02
l_0	24.87	32.95	40.00

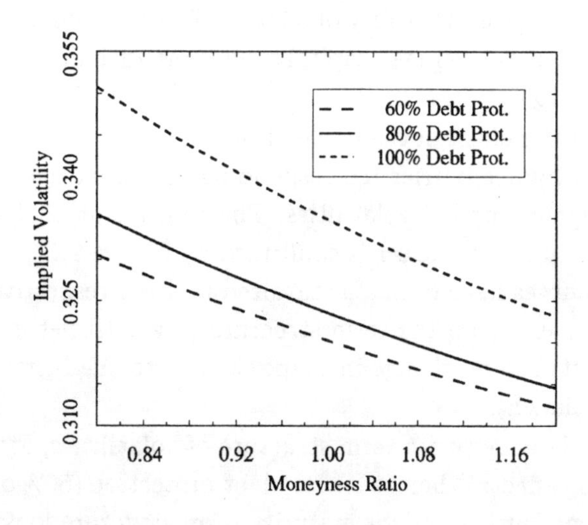

Figure 9.16: Moneyness bias in the Black–Cox (1976) capital structure for various levels of the barrier

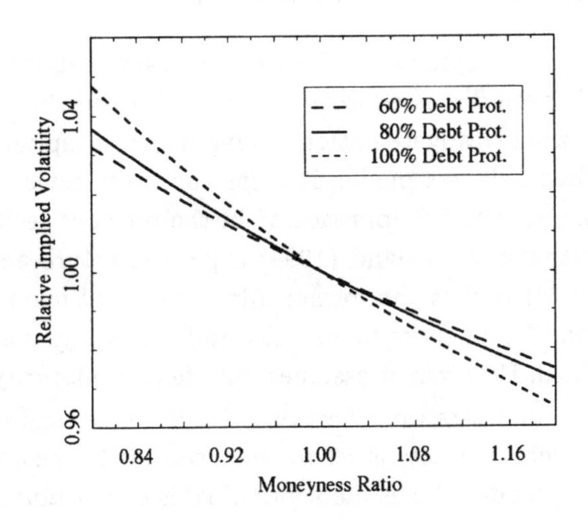

Figure 9.17: Relative implied volatilies in the Black–Cox (1976) capital structure for various levels of the barrier

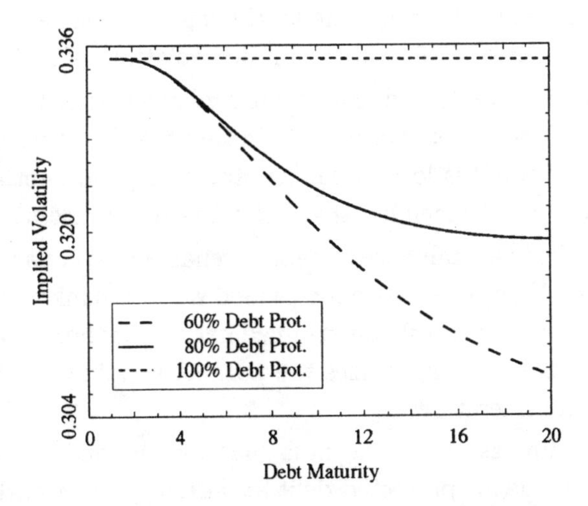

Figure 9.18: Debt-maturity term structure of volatilities in the Black–Cox (1976) capital structure for protected debt for various levels of the barrier (at-the-money option)

9.2.3 Leland (1994) / Toft and Prucyk (1997)

The option pricing implications of the Leland (1994) capital structure model have already been discussed extensively by Toft and Prucyk (1997). However, since we will investigate option pricing biases in capital structure models which extend Leland's model, it is instructive to review these biases using the same structured approach as in the previous sections. The main difference between the Leland (1994) capital structure and that of Black and Cox (1976) is that the former allows for (continuous) coupon payments, taxes, and for an endogenously determined (equity-maximizing) bankruptcy threshold. However, it assumes infinite debt maturity.

The base case parameters from Section 9.2.2 are used again with the following changes: Debt maturity is set to ∞ (according to the assumption of the Leland (1994) model). Thus, debt principal is only a *notional* principal and is never repaid. The continuous coupon is set at $C = rD$. The corporate tax rate is $\zeta = 0.35$, the fraction of bankruptcy costs is $\varphi^K = 0.1$, and the (equity-optimizing) barrier is determined endogenously. For the base case, this implies a debt principal of 47.11, an endogenous barrier at $L = 21.62$, and a payout rate of $\beta = 4.53\%$. This yields an equity value of 69.61 (which is greater than 60 due to the value of the tax shield), an at-the-money call price of 4.35, and an implied volatility of 28.53%.

The endogenously determined, equity-maximizing barrier $L = 21.62$ implies that 41.31% of total debt is protected: In the case of default, $\varphi^K = 10\%$ of firm value ($V_\tau = 21.62$) is lost to bankruptcy costs, and debtholders receive the remaining 90% (remember that Leland assumes absolute priority).

We want to look at the influence of debt covenants by comparing pricing biases for firms with unprotected debt (which declare bankruptcy at the equity-maximizing, endogenously determined barrier) to those of firms with protected debt. Table 9.3 summarizes the parameter values of the capital structures under consideration.

Figure 9.19 compares the moneyness bias for the endogenous bankruptcy level with those for protected debt at various protection levels. We see that, similar to the corresponding analysis in the previous section, implied volatilities are higher for protected debt. Moreover, the sneer is steeper for protected debt compared to unprotected debt, and the deviation increases with the protection level. This is confirmed by Figure 9.20, where all curves have been standardized by the respective at-the-money implied

Table 9.3: Parameter Values for Leveraged Capital Structures in the Leland (1994) Model for Various Debt Protection Levels

This table shows parameter values for 40 percent leveraged capital structures. Leverage is defined as $D(\cdot)/(D(\cdot)+E(\cdot))$, where $E(\cdot)$ is the market value of equity, and $D(\cdot)$ is the market value of debt. The parameters $V = 100$, $r = 0.08$, $\sigma = 0.2$, $\zeta = 0.35$, and $\varphi^K = 10\%$ are identical for all capital structures. The (equity-maximizing) barrier is determined endogenously and compared with debt protection levels of 60 and 100%, respectively. The asset payout rate β is chosen to imply a 3% annualized dividend yield on equity over the next three months.

	Debt protection		
	41.31%	60%	100%
C	3.77	3.79	3.58
D	47.11	47.32	44.78
β (%)	4.53	4.53	4.34
L	21.61	31.54	49.76
$E(\cdot)$	69.61	69.20	67.18
$D(\cdot)$	46.41	46.13	44.78

volatilities.

As far as the term structure of volatilities is concerned, Figure 9.21 shows that, while the term structure is downward-sloping for endogenous bankruptcy and low levels of debt protection, it is upward-sloping for fully protected debt (and high levels of debt protection). Overall, the extent of this effect remains small.

9.2.4 (Restricted) Leland (1998) / Model 5

The restricted version of the Leland (1998) capital structure that we presented in Section 5.12 extends the Leland (1994) model for finite debt maturity and debt issuance costs. We will focus on the effects of finite debt maturity here. The differences to the effects of debt maturity discussed for the Merton (1974) capital structure in Section 9.2.1 are that here, we have continuous coupons, taxes, intermediate bankruptcy, and bankruptcy costs. To isolate the effect of debt maturity, we assume that debt issuance costs k_1 and k_2 are equal to zero. The other parameters are the same as in the previous section, with the exception of debt maturity. In the Leland (1998)

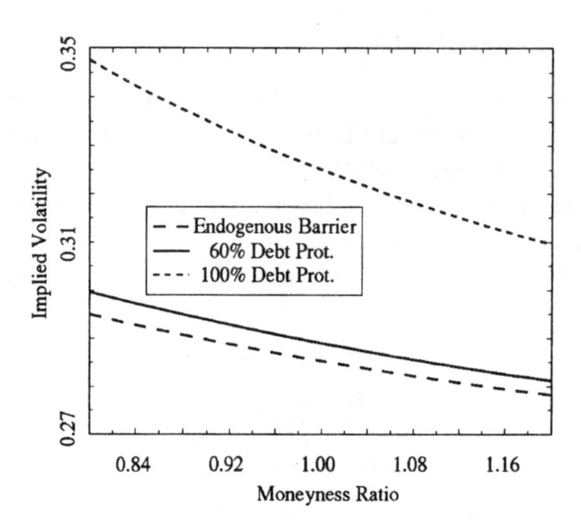

Figure 9.19: Moneyness bias in the Leland (1994) capital structure: Endogenous barrier vs. protected debt

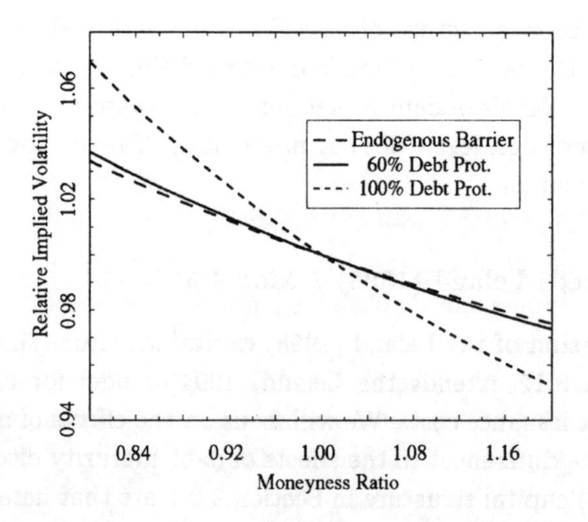

Figure 9.20: Relative implied volatilies in the Leland (1994) capital structure: Endogenous barrier vs. protected debt

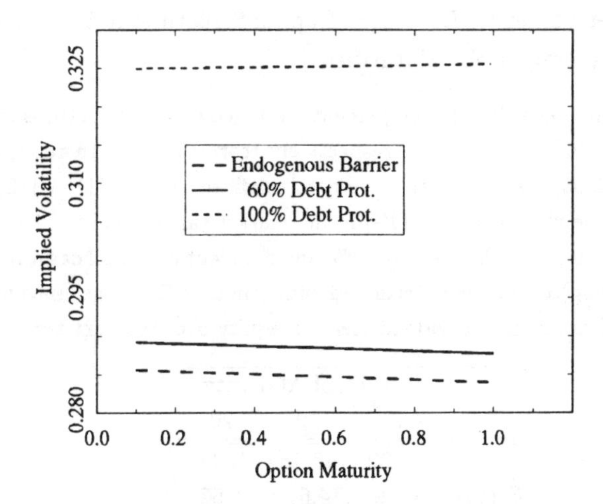

Figure 9.21: Term structure of volatilities in the Leland (1994) capital structure (at-the-money option): Endogenous barrier vs. protected debt

model, debt is issued without a fixed maturity and redeemed according to an exponential schedule. For our base case, we assume that the *average* debt maturity is 10 years, which implies a repayment rate of $m = 0.1$ (cf. equation (5.24)). Our assumption of 40% leverage and 3% dividend yield on equity implies an initial debt principal $D^{in} = 46.45$ and an asset payout rate $\beta = 4.51\%$. This yields a market value for equity of $E(\cdot) = 69.32$, an at-the-money call price of 3.72, and an implied volatility of 29.08%.

To investigate effects of average debt maturity on option pricing biases, we start by determining capital structure parameters for average debt maturities of 5 and 20 years (corresponding to repayment rates of $m = 0.2$ and $m = 0.05$, respectively), the leverage at 40% and the annualized dividend yield for equity at 3%. Table 9.4 summarizes the parameter values for these capital structures. Figure 9.22 shows the effects of average debt maturity on the moneyness bias. Similar to the Black–Cox (1976) capital structure, shorter (average) debt maturities lead to higher implied volatilities. The impression that shorter debt maturities also imply steeper sneers is confirmed by Figure 9.23, where all volatility curves have been standardized by their respective at-the-money volatilities. Comparing Figure 9.23 to Figure

Table 9.4: Parameter Values for Leveraged Capital Structures in the Leland (1998) Model for Various Average Debt Maturities

This table shows parameter values for 40 percent leveraged capital structures and average debt maturities of 5, 10, and 20 years, respectively (corresponding to repayment rates of $m = 0.2, 0.1$, and 0.05, respectively). Leverage is defined as $D(\cdot)/(D(\cdot) + E(\cdot))$, where $E(\cdot)$ is the market value of equity, and $D(\cdot)$ is the market value of debt. The parameters $V = 100$, $r = 0.08$, $\sigma = 0.2$, $\zeta = 0.35$, $\varphi^K = 10\%$, and are identical for all capital structures. The (equity-maximizing) barrier is determined endogenously. The asset payout rate β is chosen to imply a 3% annualized dividend yield on equity over the next three months.

	Av. Debt Maturity		
	5	10	20
D	46.22	46.45	46.66
β (%)	4.49	4.51	4.52
L	28.37	25.75	23.71
$E(\cdot)$	69.16	69.32	69.45
$D(\cdot)$	46.11	46.22	46.30

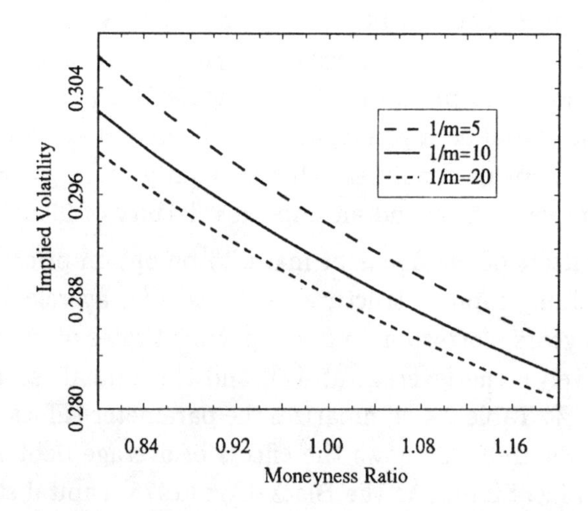

Figure 9.22: Moneyness bias in the restricted Leland (1998) capital structure for various average debt maturities $1/m$

9.8, we find that the extent of this effect is smaller here.

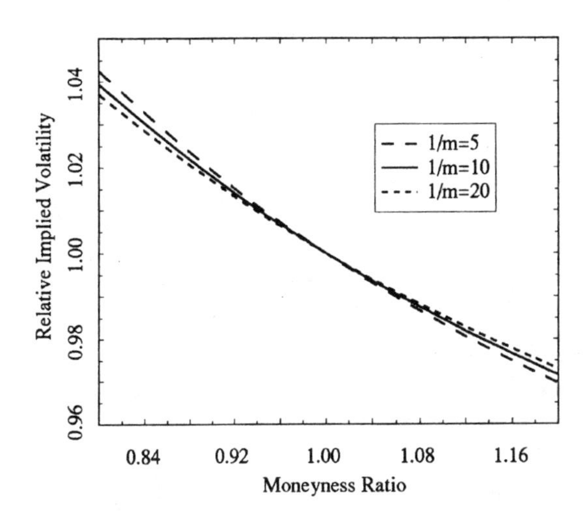

Figure 9.23: Relative implied volatilities in the restricted Leland (1998) capital structure for various average debt maturities $1/m$

Figure 9.24 shows that, unlike in the other models with finite debt maturity discussed in previous sections, the term structure of volatilities is downward-sloping for all cases considered. Again, the extent of this effect is very small.

A comparison of Figure 9.25 to Figures 9.11 and 9.15 (the corresponding figures for the Merton (1974) and Black–Cox (1976) capital structures) shows that the debt-maturity term structure for the restricted Leland (1998) capital structure behaves qualitatively differently: It is a convex function across the range of (average) debt maturities considered here. In contrast, for the other two models, it starts out concave for low debt maturities, and then turns to convex for higher debt maturities.

9.2.5 Leland and Toft (1996) / Model 4

The only extension from the Leland (1994) to the Leland and Toft (1996) model is for finite debt maturity. Due to this finite debt maturity, coupon and principal are now effectively "decoupled" (unlike in the case of infinite

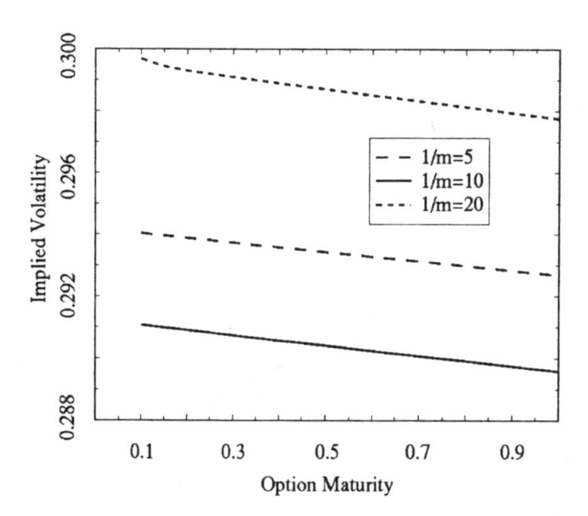

Figure 9.24: Term structure of volatilities in the restricted Leland (1998) capital structure for various average debt maturities $1/m$

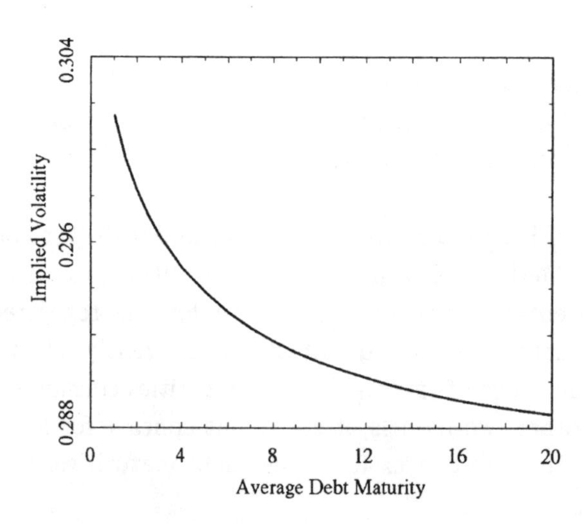

Figure 9.25: Debt-maturity term structure of volatilities in the restricted Leland (1998) capital structure for various average debt maturities $1/m$ (at-the-money option)

maturity debt, where the choice of the coupon determines the amount of principal and vice versa). Thus, the model allows for an investigation of possible effects of coupon size on option pricing biases.

The original base case maturity of $T_2 = 10$ is adopted again. All other capital structure parameters are the same as in the previous section, in particular, for the base case, we retain the assumption that $C = rD$. The (equity-optimal) barrier is determined endogenously. 40% leverage and 3% annualized dividend yield on equity imply a debt principal of $D = 46.21$ (which, by definition, corresponds to a coupon of 8%), a barrier of $L = 28.16$, and a payout rate $\beta = 4.50\%$. This yields an equity value of $E(\cdot) = 69.16$ (which, again, is greater than 60 due to the value of the tax shield), an at-the-money call price of 4.43, and an implied volatility of 29.37%.

To investigate possible effects of coupon size on option pricing biases, we start by determining capital structure parameters for 4% and 12% coupons (in addition to the base case of 8%, keeping the leverage at 40% and the annualized dividend yield for equity at 3%). Table 9.5 summarizes the parameter values for these capital structures. Figure 9.26 shows the effects

Table 9.5: Parameter Values for Leveraged Capital Structures in the Leland/Toft (1996) Model for Various Coupon Levels

This table shows parameter values for 40 percent leveraged capital structures and coupon levels of 4, 8, and 12%, respectively. Leverage is defined as $D(\cdot)/(D(\cdot)+E(\cdot))$, where $E(\cdot)$ is the market value of equity, and $D(\cdot)$ is the market value of debt. The parameters $V = 100$, $r = 0.08$, $\sigma = 0.2$, $\zeta = 0.35$, $\varphi^K = 10\%$, and $T_2 = 10$ are identical for all capital structures. The (equity-maximizing) barrier is determined endogenously. The asset payout rate β is chosen to imply a 3% annualized dividend yield on equity over the next three months.

	Coupon size		
	4%	8%	12%
D	51.49	46.21	41.93
β (%)	4.74	4.50	4.32
L	28.74	28.16	27.61

of coupon size on the moneyness bias: Although coupon size has a marked effect on the level of implied volatilities, the shape of the smile is hardly affected. The standardized curves can hardly be distinguished visually from each other (results not shown).

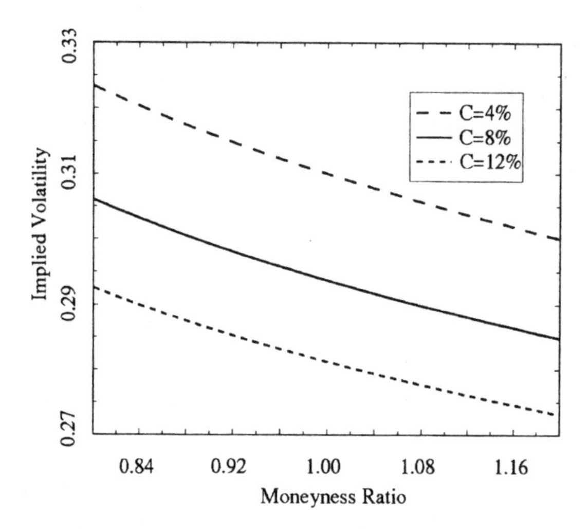

Figure 9.26: Moneyness bias in the Leland and Toft (1996) capital structure for various coupon levels

The effect of coupon size on the term structure of volatilities is illustrated in Figure 9.27. Interestingly enough, higher coupon levels lead to lower implied volatilities. As far as the term structure of volatilities is concerned, the already small effect found in previous sections is further dampened by coupons above the risk-free interest rate (in practice, most corporate bonds offer coupons above the risk-free rate). This can be better seen in Figure 9.28, where all curves have been standardized by the implied volatility of the respective option with a maturity of $T_1 = 0.5$: The lower the coupon rate, the higher the (absolute) slope of the sneer. However, in absolute terms, this effect is very small.

9.2.6 Ericsson and Reneby (2001) / Model 6

The Ericsson–Reneby (2001) model extends the Leland (1994) capital structure for exponentially increasing debt and deviations from absolute priority. Different rules of dividing the assets among securityholders in case of bankruptcy hardly affect equity options. Their effect is mostly on debt and debt derivatives, especially for firms already close to bankruptcy (results not shown).

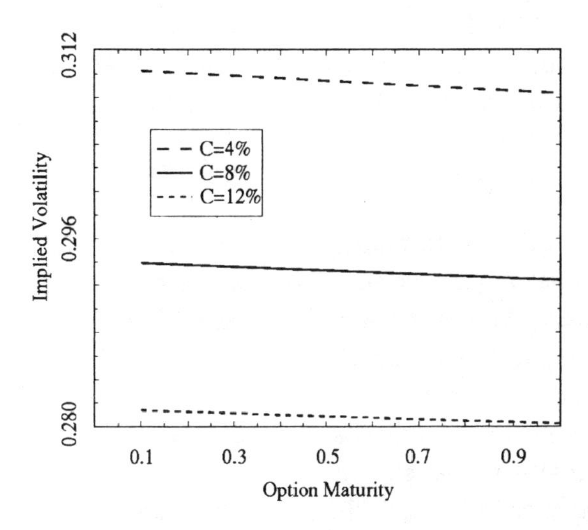

Figure 9.27: Term structure of implied volatilities in the Leland and Toft (1996) capital structure for various coupon levels (at-the-money option)

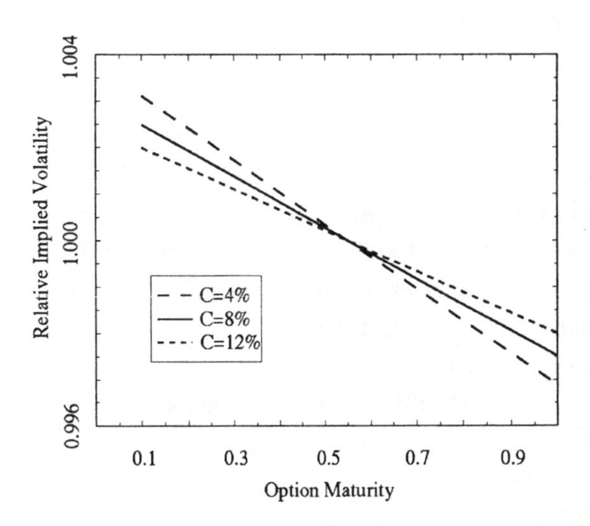

Figure 9.28: Term structure of relative implied volatilities in the Leland and Toft (1996) capital structure for various coupon levels (at-the-money option)

The effect of the growth rate of debt will be investigated in Section 9.2.7 for Model A / Model 7, which can be viewed as an extension of the Ericsson–Reneby (2001) model for finite debt maturity. For this reason, we do not provide separate numerical results for options in this framework.

9.2.7 Model A / Model 7

In this section, we investigate numerically the effects of the growth rate of debt within Model A (the extended Leland/Toft capital structure) on option pricing biases. We use the same base case parameters as in Section 9.2.5, with the understanding that the values of C and D are the coupon and principal of the bond that will mature in the next time instant.

Our results extend those of Toft and Prucyk (1997) along two dimensions: finite debt maturity and exponentially increasing debt. Since we use the same parameters as in Toft and Prucyk (1997), the results are directly comparable: Our model converges to theirs for $\nu = 0$ and $T_2 \rightarrow \infty$.[1] For $\nu = 0$ and $T_2 < \infty$, we obtain option prices within the Leland and Toft (1996) capital structure model (cf. Section 8.3). For $T = \infty$ and $\nu \geq 0$, our model converges to the Ericsson and Reneby (2001) model.

For various combinations of debt growth rate, ν, and debt maturity, T_2, we determine parameter values for 16 different capital structures, each with a leverage of 40%. Debt is assumed to be unprotected, i.e., the bankruptcy-triggering barrier is determined endogenously. As was to be expected, this endogenous barrier turns out to grow at rate ν (so we set $\rho = \nu$ in our calculations).

Parameter values for these capital structures are shown in Table 9.6. Comparing our result for $\nu = 0$ and $T_2 = 1000$ to that of Toft and Prucyk (1997), we see that the differences in parameter values are very small. Qualitatively, debt with "realistic" maturities ($T_2 = 5, 10, 15$) behaves differently from infinite (or very large) maturity debt for positive growth rates. Whereas Leland and Toft (1996, p. 998) find that the optimal (firm value maximizing) debt maturity – for unprotected debt with constant principal – is ∞, we get a different picture in the case of increasing debt: Market

[1]Numerically, we run into problems when plugging in very large values for debt maturity ($T_2 >> 1000$) because of the factor $e^{\nu T_2}$ in some of the pricing formulae (which then may exceed machine infinity). For this reason, we choose $T_2 = 1000$ as the longest debt maturity in our numerical analysis. As shown in Table 9.6, the differences in parameter values for $T_2 = 1000$ vs. $T_2 \rightarrow \infty$ are very small.

Table 9.6: Parameter Values for Leveraged Capital Structures

This table shows parameter values for 40 percent leveraged capital structures. Leverage is defined as $D(\cdot)/(D(\cdot)+E(\cdot))$, where $E(\cdot)$ is the market value of equity, and $D(\cdot)$ is the market value of all outstanding debt. These values are computed using equations (5.30) and (5.28). The parameters $V = 100$, $r = 0.08$, $\sigma = 0.2$, $\varphi^K = 0.1$, $\varphi^D = 0.9$, and $\zeta = 0.35$ are identical for all capital structures. The asset liquidation rate, β, is determined so that the dividend payable to equity holders during the next three months is equivalent to a 3 percent dividend yield on equity. Bankruptcy is determined endogenously at the level which maximizes equity value, subject to its limited liability. For easy comparison, the last column shows the corresponding result of Toft and Prucyk (1997, p. 1162).

		Debt Maturity				
Debt Growth Rate		5	10	15	1000	∞ (TP97)
	C	3.68	3.70	3.71	3.77	3.77
	β (%)	4.47	4.50	4.52	4.53	4.54
$\nu = 0$	l_0	31.49	28.16	26.50	21.72	21.61
	$E(\cdot)$	68.95	69.16	69.27	69.60	69.61
	$D(\cdot)$	45.96	46.21	46.18	46.41	46.41
	C	3.59	3.41	3.27	2.48	
	β (%)	3.34	3.45	3.51	3.71	
$\nu = 3\%$	l_0	32.28	28.20	26.02	18.86	
	$E(\cdot)$	71.96	71.84	71.58	69.32	
	$D(\cdot)$	47.97	47.90	47.72	46.20	
	C	3.58	3.19	2.89	1.36	
	β (%)	2.18	2.40	2.55	3.21	
$\nu = 6\%$	l_0	33.91	28.96	26.10	14.20	
	$E(\cdot)$	76.74	76.04	75.10	68.33	
	$D(\cdot)$	51.16	50.69	50.07	45.55	
	C	3.65	2.97	2.50	0.60	
	β (%)	0.96	1.35	1.63	2.94	
$\nu = 9\%$	l_0	36.58	30.30	26.47	8.44	
	$E(\cdot)$	83.77	80.39	77.70	66.26	
	$D(\cdot)$	55.85	53.89	51.79	44.17	

values of both equity and debt are decreasing in debt maturity. Positive growth rates for debt, however, are desirable, because – as can be seen from Table 9.6 – they lead to higher market values for both equity and debt for realistic values of T_2. Thus, a debt growth rate of $\gamma = 0$ (as assumed by Leland and Toft (1996)) turns out to be suboptimal. Using our model, their analysis of optimally levered firms could be extended for the dimension of debt growth rate.

However, our main interest here lies with option pricing. To facilitate comparison with previous results, we compute prices for options on equity in the capital structures described in Table 9.6. These option prices are then converted into implied volatilities. To facilitate comparisons across strike prices, we divide the implied volatilities by the implied volatility of the respective at-the-money option. Table 9.7 shows the resulting matrices of implied Black–Scholes volatilities.

It illustrates a number of interesting results: First, for options in the Leland and Toft (1996) capital structure ($\gamma = 0$), the implied volatility structure is more negatively skewed for shorter debt maturities. Second, for "reasonable" debt maturities, the negative skew increases with higher debt growth rates, whereas for $T_2 = 1000$ (which may be regarded as infinite debt maturity for practical purposes), the negative skew decreases with higher debt growth rates. We further observe that, compared to the skews in Toft and Prucyk (1997) (which seem to underestimate observed volatility skews), our skews are more pronounced and therefore closer to empirically observed pricing biases.[2]

We summarize these observations by formulating the following empirically testable hypotheses:

> *Hypothesis 1:* The volatility skew is negatively related to debt maturity: Firms with shorter debt maturities show steeper sneers compared to firms with longer debt maturities.
> *Hypothesis 2:* For reasonable values of debt maturity and debt growth rate, the volatility skew is positively related to the growth rate of debt: Firms with higher debt growth rates show steeper sneers compared to firms with lower growth rates.

[2]For the extent of empirically observed biases, see e.g. Dumas, Fleming, and Whaley (1998, p. 2063).

Table 9.7: Relative Implied Black–Scholes Volatilities as Functions of Moneyness and Debt Maturity

This table shows relative implied Black–Scholes volatilities as functions of moneyness (defined in this table as strike divided by equity price) and debt maturity for the capital structures described in Table 9.6. The option maturity is $T_1 = 0.25$, the other input parameters are as shown in Table 9.6. The relative implied Black–Scholes volatility is defined as the volatility implied from the Black–Scholes formula when the option price is determined by the proposed model for options on leveraged equity, divided by the implied volatility for the corresponding at-the-money option. For easy comparison, the last column shows the corresponding results of Toft and Prucyk (1997, p. 1162).

		Debt Maturity				
	Moneyness	5	10	15	1000	∞ (TP97)
	80%	1.047	1.042	1.040	1.034	1.034
$\nu = 0$	90%	1.021	1.019	1.018	1.016	1.015
	110%	0.982	0.984	0.985	0.987	0.987
	120%	0.967	0.970	0.971	0.975	0.975
	80%	1.053	1.045	1.042	1.031	
$\nu = 3\%$	90%	1.024	1.021	1.019	1.014	
	110%	0.980	0.982	0.984	0.988	
	120%	0.962	0.967	0.970	0.977	
	80%	1.061	1.051	1.046	1.026	
$\nu = 6\%$	90%	1.028	1.024	1.021	1.012	
	110%	0.976	0.980	0.982	0.990	
	120%	0.956	0.962	0.966	0.981	
	80%	1.062	1.052	1.046	1.017	
$\nu = 9\%$	90%	1.029	1.024	1.021	1.008	
	110%	0.975	0.979	0.981	0.993	
	120%	0.953	0.961	0.965	0.987	

Note that to make the results comparable, the corresponding values from Toft and Prucyk (1997, Table II) have been divided by 0.999, the relative volatility given there for the *three month* at-the-money option. By construction, table entries for at-the-money options (suppressed) would be 1.000 throughout.

9.2.8 Conclusions

From the numerical analysis of pure debt-equity capital structures in the previous sections, we draw the following main conclusions:

1. The asymmetric volatility smile observed in equity option markets occurs in all firm value based pricing models considered. However, for all capital structure models considered, the assumption of *unprotected* debt leads to asymmetries which are less pronounced than those observed in practice. For reasonable growth rates of debt, Model 7 seems to be the one that – even for unprotected debt – produces smiles which are the closest to realistic magnitudes among all models considered here.

2. Structural credit risk (i.e., the possibility of changes in a firm's capital structure) is definitely *not* a major factor behind the observed *term structure* of volatilities. The common explanation of varying equity volatility seems to contribute much more to this phenomenon. For a constant volatility of (V_t), however, we find for those models that are closest to reality that the term structure is invariably slightly downward-sloping. So, for empirical testing, we would predict that – under the assumption of a "symmetric", mean-reverting stochastic volatility – we should observe downward-sloping term structures more often than upward-sloping term structures.

3. The debt-maturity term structure of volatilities seems to be a very interesting phenomenon that has not received any attention so far in the literature. Given that changes in a firm's (average) debt maturity are not uncommon in practice, we feel that this aspect deserves much more attention, both in future research and in practice.

9.3 Capital Structure Models with Convertibles

The setup used by Hanke and Pötzelberger (2002) for pricing options in the presence of warrants in a firm's capital structure focuses on the effect of warrant *issuance* on option prices. Therefore, it is better suited to investigate effects of *changes* in a firm's capital structure (rather than a static analysis), and we will defer numerical results for this model to Chapter 10. Here, we will investigate option pricing biases within the Ingersoll (1977a) capital structure model(s), featuring callable and non-callable convertible discount bonds. Option pricing extensions to these capital structures have been derived in Section 8.2.

In the following analysis of deviations between option prices in these models and those from the Black–Scholes formula, we use the same base case parameters as in Section 9.2.1: $V = 100$, $r = 0.08$, $\sigma = 0.2$, $T_2 = 10$, and a leverage of 40%. In addition, we use a conversion rate of $\gamma = 0.15$.

9.3.1 Ingersoll (1977a): Convertible Discount Bonds / Model 2

For the case of non-callable convertible debt, our base case parameters imply a debt principal of 91.36 and a market value of equity of 60 (no taxes). The corresponding at-the-money call price is 4.36, and the implied volatility 31.55%. To investigate the impact of the conversion rate γ on the smile, we contrast the cases of $\gamma = 0$ (which corresponds to the Merton (1974) capital structure) to $\gamma = 0.15$ (our base case) and $\gamma = 0.3$. We get the implied volatility smile in Figure 9.29. Clearly, a higher conversion rate leads to lower implied volatilities, whereby the change in implied volatility when changing γ from 0 to 0.15 is much smaller than when changing γ from 0.15 to 0.3. The shape of the smile, however, is hardly affected.

The same effect can be seen from Figure 9.30, which shows the term structure of volatilities for this model. The effect of a higher conversion rate is a reduction in implied volatility, but no change in the shape of the (upward-sloping) term structure.

In contrast to the previous analyses, the debt maturity term structure of volatilities shows qualitative differences depending on the level of γ, as can be seen from Figure 9.31: Whereas for low levels of γ, the curve starts out convex for short debt maturities and turns concave for higher debt maturities, it is concave throughout for $\gamma = 0.3$.

9.3.2 Ingersoll (1977a): Callable Convertible Discount Bonds / Model 3

For the case of callable convertible debt, we assume that the call price grows exponentially at rate $\rho = r$, which implies an up-barrier at $l_t = De^{-r(T-t)}/\gamma$. Keeping all the other parameters at their previous levels, we find that the callability feature changes the debt principal to $D = 92.29$, the at-the-

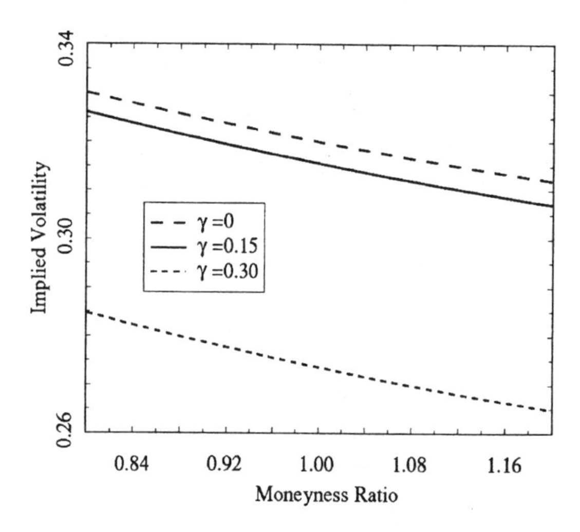

Figure 9.29: Moneyness bias in the Ingersoll (1977a) capital structure (base case, non-callable convertible debt) for various levels of the conversion rate γ

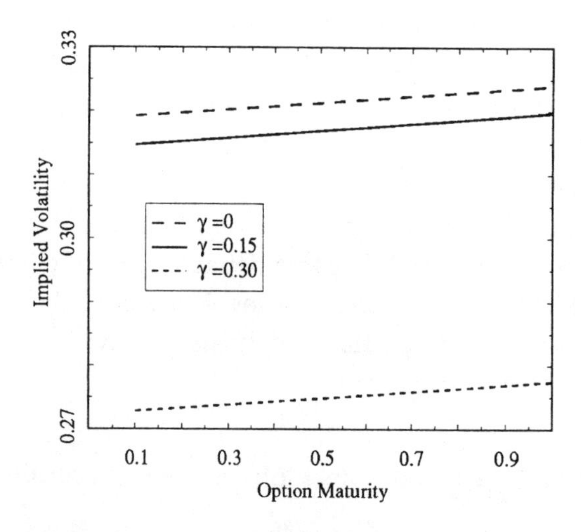

Figure 9.30: Term structure of volatilities in the Ingersoll (1977a) capital structure (base case, non-callable convertible debt) for various levels of the conversion rate γ

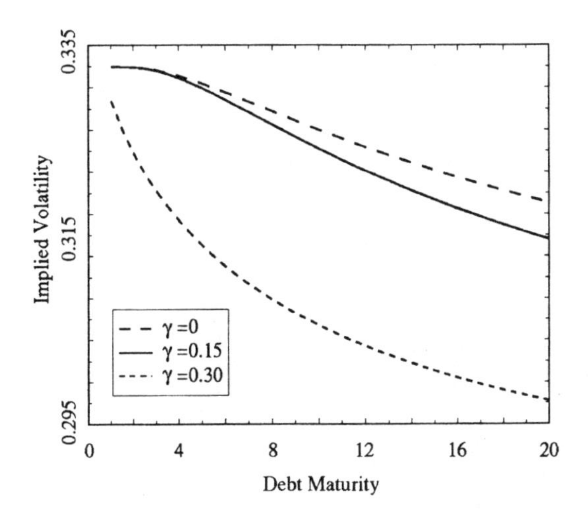

Figure 9.31: Debt maturity term structure of volatilities in the Ingersoll (1977a) capital structure (base case, non-callable convertible debt) for various levels of the conversion rate γ

money call price to 3.89, and the implied volatility to 31.98%. The level of the barrier at time 0 is $\mathfrak{l}_0 = 276.45$.

Figure 9.32 shows that the callability feature drastically reduces the impact of the conversion rate: The smiles for $\gamma = 0$ and $\gamma = 0.15$ cannot be distinguished visually, and the difference in implied volatility between $\gamma = 0$ and $\gamma = 0.3$ becomes very small.

The picture for the debt-maturity term structure of volatilities is similar: Figure 9.33 shows that the curves for $\gamma = 0$ and $\gamma = 0.15$ coincide, and only for longer debt maturities do we find a significant impact of the conversion rate on the level of implied volatilities. The qualitative differences observed in the non-callable case (cf. Figure 9.31), however, have disappeared.

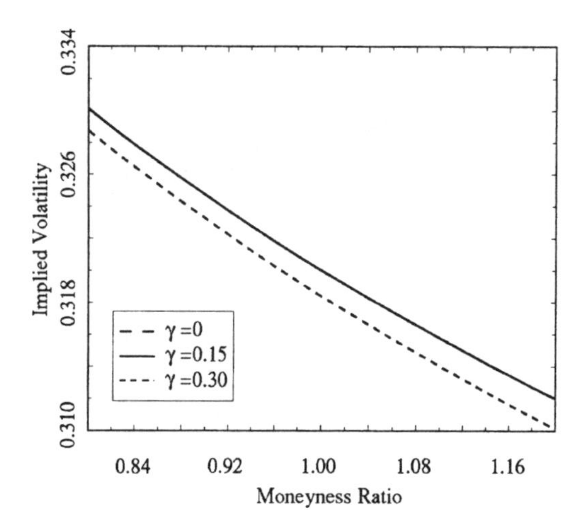

Figure 9.32: Moneyness bias in the Ingersoll (1977a) capital structure (base case, callable convertible debt) for various levels of the conversion rate γ

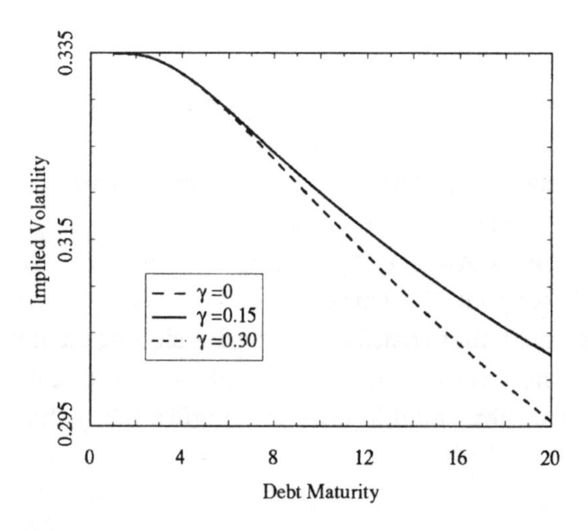

Figure 9.33: Debt-maturity term structure of volatilities in the Ingersoll (1977a) capital structure (base case, callable convertible debt) for various levels of the conversion rate γ

Chapter 10

Option Pricing Effects of Changes in a Firm's Capital Structure

In this chapter, we will use numerical examples to illustrate another implication of firm value based option pricing: Changes in model prices of options induced by changes in the firm's capital structure. Intuitively, and with the help of the static results presented in Chapter 9, it should be obvious that an increase in debt, e.g., leads to higher option prices due to an increased equity volatility. However, there exists almost no literature on this topic as far as the quantification of this effect is concerned.

For rather simple capital structures, Bensoussan, Crouhy, and Galai (1994) approximate equity volatility in a firm value based framework *without* intermediate default (i.e., in a setting which is much simpler than the one presented in this book). They note that

> "instantaneous equity volatilily is a solution of a partial differential equation similar to Black–Scholes', although it is non-linear and in general does not have any analytic solution".

However, for exogenous barriers, the derivatives of the building blocks presented in Chapter 5 can be expressed in closed form (see, e.g., Ericsson (1997, pp. 32ff.)). Using the well-known relation

$$\sigma_E = \frac{\partial E}{\partial V} \frac{V}{E},$$

equity volatility is available in closed form for all these models.[1] However, although changes in equity volatility are the most important factor behind

[1]When the barrier is determined endogenously, things become more difficult, and a closed-form solution is, in general, no longer available.

option price effects induced by changes in capital structure, skewness and kurtosis of the equity return distribution also change and affect option prices. Therefore, our approach (which, implicitly, takes into account all changes in the equity return distribution) is clearly preferable to that taken by Bensoussan, Crouhy, and Galai (1994).

To the best of our knowledge, the first paper directly quantifying option price effects induced by capital structure changes within a "true" firm value-based framework is Hanke and Pötzelberger (2002) (described in Section 7.3). Generally speaking, the number of interesting cases is too large to cover them all here. For this reason, we confine ourselves to some illustrative examples within the "richest" model available (Model A/Model 7). As an example of capital structure effects of changes "within equity", we close the chapter with a review of the results from Hanke and Pötzelberger (2002).

10.1 Changes within Model 7

We will examine the effects of the following capital structure changes:

1. Changes in the level of debt.
2. Changes in the growth rate of debt.
3. Changes in debt maturity.
4. Changes in the level of debt protection.

Whereas the first three cases are "true" capital structure changes, the fourth falls into the more general category of "changes in the firm's financial relations".

10.1.1 Changes in the Level of Debt

We use the same set of base case parameters as in Section 9.2.7: $V = 100$, $r = 0.08$, $\sigma = 0.2$, $T_1 = 0.25$, $T_2 = 10$, $C = rD$, where we remember that C applies to the bond which matures in the next time instant. The corporate tax rate is set at $\zeta = 0.35$, the barrier grows at the same rate as debt ($\rho = \nu$). The growth rate of debt is set to $\nu = 0.03$. The fractions of asset value in case of default received by debtholders / lost to bankruptcy costs are $\varphi^D = 0.9$ and $\varphi^K = 0.1$. Debt principal D and asset payout rate β are chosen such that leverage is 40% and the (annualized) dividend yield on equity over the option's lifetime, is $q = 0.03$. This implies a debt principal of $D = 42.63$, and a payout rate of 3.45% (cf. Table 9.6).

Using these parameters, we get an equity value of 71.84, and a corresponding at-the-money call price of 4.599. We want to investigate the effects of a change in leverage. More for mathematical convenience rather than economic reasons, we assume that any debt increase is made by increasing the principal of all outstanding bonds such that the principal of the bond maturing in the next time instant changes from D to D^{new}, and the principal of the bond with the longest maturity changes from $De^{\gamma T_2}$ to $D^{new}e^{\gamma T_2}$ (analogously for a decrease in debt). To avoid any jumps in (V_t) arising from a possible expansion (contraction) of the firm's business activities using the cash inflow (outflow) due to the change in capital structure, we assume that any cash inflow from newly issued debt is used to repurchase shares (and any cash outflow from a decrease in debt is funded by newly issued equity). This allows us to focus exclusively on effects of an increased debt-equity ratio, rather than the effect of, e.g., a debt-financed expansion. However, the strike of the option must be adjusted to reflect the share repurchase. This can be done easily using

$$K^{new} = K\frac{E^{new}}{E},$$

where E^{new} denotes the value of equity after the change in capital structure as described above. In addition, to ensure comparability of the results, we standardize the resulting option prices by the corresponding equity values. In this way, we compare options on a fraction of equity worth 1 unit of currency both before and after debt issuance and share repurchase, and avoid possible misinterpretations resulting from scaling effects.

Similar to Hanke and Pötzelberger (2002), we assume that calls with moneyness ratios of 0.9/1/1.1 are already outstanding at the time of the capital structure change. Table 10.1 summarizes the resulting call price changes. We find that prices of out-of-the-money calls are most affected by a change in leverage. Even a modest increase in leverage from 40% to 50% leads to an increase in the option's price by almost 40%, whereas the same change in capital structure causes a price increase for the in-the-money option of only around 7%.

Table 10.1: Change in call price (in %) due to change in leverage (Model A/Model 7).

	New leverage (in %)			
Moneyness Ratio	20	30	50	60
0.9	-6.77	-4.26	7.15	18.82
1.0	-19.80	-12.00	18.36	46.15
1.1	-39.09	-24.20	39.27	102.14

The underlying is a fraction of equity worth one unit of currency both before and after the debt increase (i.e., the terms of the option are assumed to be adjusted for the share repurchase). The base case leverage is 40%.

10.1.2 Changes in the Growth Rate of Debt

The main argument for the assumption of an increase in debt over time (cf. the discussion in Section 5.13) was that otherwise, if firms grow over time (on average), their debt-equity ratio would go to zero. In this section, we assume that the firm wants to keep its debt-to-equity ratio roughly constant over time. Therefore, it chooses a growth rate of debt that matches the growth rate of (V_t): $\nu = \mu_V$. We assume that the current value of $\nu = 0.03$ is justified by an average firm value growth rate of μ_V=3%. Should growth expectations change, ν would be adjusted accordingly. The optimal barrier is determined endogenously and grows at a rate of $\rho = \nu$ (cf. Section 9.2.7).

For the exact adjustment procedure, we make again some mathematically convenient assumptions. In case of an increase in ν, e.g., we assume that retired debt will be replaced by new debt with principal equal to $\exp(\nu^{new})$ times the principal of the retired debt. In addition, *all* currently outstanding debt will be increased to correspond to the new growth rate. E.g., the principal of the bond with the longest maturity is increased from $e^{\nu T_2}D$ to $e^{\nu^{new} T_2}D$. The additional cash raised from increasing currently outstanding debt is used to repurchase shares. In case of a decrease in ν, the retired debt is replaced by newly issued equity. The adjustments necessary to ensure comparability of option prices are the same as in the previous section.

Table 10.2 shows the effects on call prices of a change in the growth rate of debt from currently $\nu = 0.03$ to $\nu = -0.03$ (corresponds to a contracting firm), $\nu = 0$ (no growth), and $\nu = 0.06/\nu = 0.09$ (high growth).

Table 10.2: Change in call price (in %) due to change in the growth rate of debt ν (Model A/Model 7).

	New growth rate (ν^{new}, in %)			
Moneyness Ratio	-3%	0	6%	9%
0.9	0.27	0.00	1.16	4.48
1.0	0.97	0.15	2.96	11.76
1.1	2.51	0.62	5.86	25.00

The underlying is a fraction of equity worth one unit of currency both before and after the change in the growth rate of debt (i.e., the terms of the option are assumed to be adjusted for the share repurchase/issuance). The base case growth rate is 3%.

Again, out-of-the-money options are the strongest affected. Overall, however, the effect is much smaller than that of a change in leverage. Only sizeable changes in the growth rate of debt would lead to noticable option price effects.

10.1.3 Changes in Debt Maturity

A "classical" conflict of interest between borrowers and lenders is that, boldly simplified and interest rate risk aside, borrowers would like to borrow long term, whereas lenders would rather like to lend short term to reduce credit risk. Short-term debt, such as revolving lines of credit, is also known in the agency literature as a signal by the borrower of acting in the lender's best interest. The reasoning behind this is that, in the case of short-term debt, the lender has the power to refuse renewal of the line and thereby put the borrower into serious trouble.

Generally, it will be harder for a firm with higher credit risk to borrow long term than for an investment-grade firm. Thus, changes in a company's credit rating may lead to changes in average debt maturity, which in turn affects prices of options on equity.

We assume – for mathematical convenience – that any change in debt maturity may also affect currently outstanding debt. If, e.g., debt maturity is increased, this is achieved by issuing bonds with maturities up to the new maturity T_2^{new}. Conversely, if debt maturity is decreased, bonds with maturities in excess of the new maturity T_2^{new} are redeemed immediately. Any cash inflows from these transactions are used to repurchase shares, and any cash needed is raised by issuing new shares.

Table 10.3 shows percentage changes in call prices for various debt maturities, relative to the base case maturity of $T_2 = 10$. Similar to the growth

Table 10.3: Change in call price (in %) due to change in debt maturity (Model A/Model 7).

	New debt maturity		
Moneyness Ratio	5	15	20
0.9	.81	-.38	-.60
1.0	1.94	-0.92	-1.44
1.1	3.66	-1.70	-2.66

The underlying is a fraction of equity worth one unit of currency both before and after the change in the growth rate of debt (i.e., the terms of the option are assumed to be adjusted for the share repurchase/issuance). The base case maturity is 10 years.

rate of debt, a change in debt maturity has only a small effect on option prices. Again, it is more pronounced for higher moneyness ratios.

10.1.4 Changes in the Level of Debt Protection

The barrier $\mathfrak{l}_t$ on asset value is often used to model so-called *debt covenants.* The higher $\mathfrak{l}_t$, the larger the payout rate for creditors in the case of default. In the absence of such covenants, there exists a lower bound on (V_t) (we called it L_{end} in previous chapters) such that equityholders will "give up" the firm by refusing to make any further contributions.

It is plausible to assume that changes in the firm's credit rating will affect the level of the barrier: If creditworthiness falls, the firm will only be able to issue new debt if it offers creditors increased protection. We are interested in the effect of changes in the barrier on prices of equity options.

We use the same parameters as in the previous sections with the following exception: The barrier is no longer determined endogenously, but represents now debt covenants. The equity-optimal endogenous barrier is $\mathfrak{l}_0 = 28.20$, which means that only barrier values above this value are economically interesting. We will vary the endogenous barrier in the interval $[40, 60]$ and investigate its influence on option prices. For mathematical convenience, we assume that any request by creditors for additional protection affects not only new debt, but also currently outstanding debt (remember that exponential growth of the barrier was an important condition in the derivation of our valuation formulae). Contrary to the cases discussed in the previous sections, however, there will be no repurchase or issuance of new shares, and, hence, no adjustment to the terms of the option. This means that the call price will be decreased by a lower equity value (due to the higher barrier).

Table 10.4 shows – not surprisingly – that a higher barrier leads to markedly lower option prices for all moneyness ratios. Out-of-the-money options are again the most affected, although differences across moneyness ratios are less severe than in previous cases: An increase in the barrier from 28.20 to 40 leads to a decrease in the option price by 20 to 30%.

Table 10.4: Change in call price (in %) due to change in barrier level (Model A/Model 7).

	New barrier level l_0		
Moneyness Ratio / l_0	40	50	60
0.9	-20.94	-41.51	-63.45
1.0	-25.43	-48.00	-69.60
1.1	-29.29	-53.06	-73.93

The underlying is a fraction of equity worth one unit of currency both before and after the change in the growth rate of debt (i.e., the terms of the option are assumed to be adjusted for the share repurchase/issuance). The base case barrier is 28.20 (i.e., the equity-maximizing endogenous barrier).

10.2 Changes within Hanke and Pötzelberger (2002)

The framework of Hanke and Pötzelberger (2002) is particularly suitable for the investigation of the effects of warrants issuance on the prices of stock options. The change in firm value due to the cash inflow from the sale of the warrants is explicitly taken into account (via standard reinvestment assumptions, cf. Section 7.3).

Hanke and Pötzelberger (2002) investigate the effects of warrants issuance on option prices for realistic input parameters. They select 14 warrants listed in the Value Line Convertibles survey (for a more detailed description of the data, see Hanke and Pötzelberger (2002, pp. 76f.)) and examine changes in prices of nine standardized call options: These cover a grid of moneyness ratios (0.9, 1, and 1.1), as well as option maturities (1, 4 and 12 weeks). For the stock volatility before warrants issuance, they use the implied equity volatility calculated from the observed warrant price at issuance.

We demonstrate the calculations using the warrant on AVI Biopharma (symbol: AVIIW) as an example. The stock price at the time of warrant issuance was 6.5, the strike is 13.5, the dilution $m/n = 0.18$, and the warrant maturity $T = 5$ (years). Assuming a risk-free interest rate of 5%, they compute an implied equity volatility from equation (7.4) of 57%. The Black–Scholes price of an at-the-money European call option with a maturity of $T_1 - t_0 = 1/52$ (1 week) immediately before warrants issuance is 0.2079598. The price of the same option immediately after warrants issuance is computed using equation (7.5) to 0.1995675. This gives a price reduction of $1 - 0.1995675/0.2079598 = 4.04\%$.

The resulting price changes for different option maturities are given in Table 10.5. From these results, we derive the following general findings:

- For parameter settings occurring in practice, warrants issuance may lead to a loss in value for an option on the stock of up to 98% (IFLYW for the option with a remaining lifetime of one week).
- The shorter the option's remaining life, the higher the price reduction induced by warrants issuance.
- Low volatility and high dilution (relative size of warrant issue) lead to large option price decreases.
- In-the-money options are hardly affected by warrants issuance.
- A "typical" price decrease for an at-the-money option is around 10%, depending on option maturity.

Table 10.5: Price decrease in % for a standard European call option induced by warrants issuance for different call maturities (ranked by implied equity volatility)

Warrant symbol	Input parameters						Price decrease (in %) Option moneyness		
	S	x	$\hat{\sigma}$	$\frac{m}{n}$	T	$\frac{T^o}{52}$	$\frac{x^o}{S}=0.9$	$\frac{x^o}{S}=1$	$\frac{x^o}{S}=1.1$
IFLYW	5.125	6.25	15%	36%	4.98	1	0.00	13.07	98.67
						4	0.04	12.30	72.79
						12	0.68	11.19	43.98
TAMRW	4.984	6	15%	14%	5	1	0.00	5.81	81.23
						4	0.02	5.47	40.12
						12	0.24	4.98	20.79
AESPW	7.25	6.9	19%	35%	5	1	0.00	14.04	94.13
						4	0.23	13.41	60.72
						12	1.68	12.51	37.00
RTROW	7	7.5	21%	27%	5	1	0.00	10.44	81.82
						4	0.31	10.01	44.59
						12	1.63	9.39	26.40
PCTHW	2.563	4.69	29%	6%	6.65	1	0.01	1.69	13.70
						4	0.19	1.64	5.85
						12	0.48	1.56	3.55
DECTW	7.875	9	35%	36%	5	1	0.13	10.52	53.45
						4	1.75	10.26	28.47
						12	3.85	9.89	19.10
NTFYW	4.438	6.5	35%	36%	5	1	0.12	9.51	50.04
						4	1.57	9.27	26.24
						12	3.41	8.90	17.49
ADSTW	8	9	36%	35%	4.92	1	0.16	10.21	50.79
						4	1.82	9.97	27.03
						12	3.86	9.61	18.25
PSCOW	6.25	6.32	47%	59%	2.85	1	0.79	15.72	56.42
						4	4.37	15.43	33.76
						12	7.55	15.02	24.75
UVSLW	2.125	2.25	51%	19%	1.47	1	0.49	6.42	24.38
						4	2.03	6.30	13.70
						12	3.23	6.13	10.03
AVIIW	6.5	13.5	57%	18%	5	1	10.45	4.04	13.82
						4	1.50	3.97	7.94
						12	2.24	3.86	5.93
AASIW	5.125	6.5	62%	86%	5	1	1.88	14.63	42.29
						4	5.89	14.43	26.59
						12	8.65	14.14	20.61
RACNW	1.5	3.54	69%	7%	1.78	1	0.24	1.38	4.15
						4	0.59	1.35	2.51
						12	0.79	1.29	1.91
VNCIW	1.875	4	90%	45%	3.94	1	2.06	7.29	16.60
						4	4.33	7.21	11.26
						12	5.11	7.08	9.26

Chapter 11

Conclusions and Directions for Further Research

Following an introduction to the idea of firm value based pricing of equity options and a description of the necessary mathematical tools, we have presented a probabilistic firm value based pricing framework proposed by Ericsson and Reneby (1996, 1998, 2001). This framework has then been used to review the most important firm value based pricing models for corporate securities and options on equity. By considerably extending the framework through the derivation of new formulae for additional building blocks, we have shown that almost all classical models (that assume a constant risk-free interest rate) can be re-derived within this (extended) framework in a straightforward way. The modularity of this approach greatly simplifies economic interpretation of the resulting formulae.

A new capital structure model (Model A) has been presented, which can be viewed as an extension of the Leland and Toft (1996) capital structure for exponentially increasing debt. Moreover, closed-form option pricing extensions have been provided both for this model, and for a number of classical capital structure models. These formulae directly incorporate structural credit risk into the prices of equity options.

Numerical analyses of the resulting formulae have shown that structural credit risk may be a driving factor behind the existence of asymmetric volatility smiles in option markets. Moreover, the relationship between (average) debt maturity and pricing biases of equity options has been investigated. We have suggested using the term "debt-maturity term structure of volatilities" for this phenomenon, which has gone largely unnoticed in the theoretical and empirical literature.

Further research should focus on the following aspects:

- As far as the practical application of our results is concerned, the findings from our numerical analyses should be tested empirically (along the lines of Toft and Prucyk (1997)).

- Then, an estimation of the economically richest models (Model A/Model 7) and a comparison with observed market prices of primary/derivative securities would be very interesting.

- On the theoretical side, an extension to stochastic interest rates would be desirable. The most promising path seems to be via an extension of the Longstaff and Schwartz (1995) model. Complications that may arise are the time-dependence of martingale measures and the non-existence of "genuinely closed-form" solutions. However, it should be possible to derive "quasi-closed-form" solutions for option prices similar to the debt valuation formulae in Longstaff and Schwartz (1995).

- Whereas, in this book, we have assumed that all market participants are fully informed about all the facts which are relevant for pricing, an investigation of the option pricing problem under *asymmetric information* would also be very interesting.

Bibliography

Baxter, M., and A. Rennie, 1996, *Financial Calculus*. Cambridge University Press, Cambridge.

Bensoussan, A., M. Crouhy, and D. Galai, 1994, "Stochastic Equity Volatility Related to the Leverage Effect," *Applied Mathematical Finance*, 1, 63–85.

Björk, T., 1998, *Arbitrage Theory in Continuous Time*. Oxford University Press, Oxford.

Black, F., and J. Cox, 1976, "Valuing Corporate Securities: Some Effects of Bond Indenture Provisions," *The Journal of Finance*, 31, 351–367.

Black, F., and M. Scholes, 1973, "The Pricing of Options and Corporate Liabilities," *Journal of Political Economy*, 81, 637–654.

Brennan, M. J., and E. Schwartz, 1977, "Convertible Bonds: Valuation and Optimal Strategies for Call and Conversion," *The Journal of Finance*, 32, 1699–1715.

———, 1978, "Corporate Income Taxes, Valuation, and the Problem of Optimal Capital Structure," *Journal of Business*, 51, 103–114.

———, 1980, "Analyzing Convertible Bonds," *Journal of Financial and Quantitative Analysis*, 15, 907–929.

Christensen, P., C. Flor, D. Lando, and K. Miltersen, 2002, "Dynamic Capital Structure with Callable Debt and Debt Renegotiation," working paper, Dept. of Accounting, Finance, and Law, University of Southern Denmark, Odense.

Constantinides, G., 1984, "Warrant Exercise and Bond Conversion in Competitive Markets," *Journal of Financial Economics*, 13, 371–397.

Cox, J., J. Ingersoll, and S. Ross, 1985, "A Theory of the Term Structure of Interest Rates," *Econometrica*, 53, 385–407.

Cox, J., S. Ross, and M. Rubinstein, 1979, "Option Pricing: A Simplified Approach," *Journal of Financial Economics*, 7, 229–263.

Dalang, R., A. Morton, and W. Willinger, 1990, "Equivalent Martingale Measures and No-Arbitrage in Stochastic Securities Market Models," *Stochastics and Stochastic Reports*, 29, 185–201.

Delbaen, F., 1992, "Representing Martingale Measures When Asset Prices Are Continuous and Bounded," *Mathematical Finance*, 2, 107–130.

Delbaen, F., and W. Schachermayer, 1994a, "Arbitrage and Free Lunch with Bounded Risk for Unbounded Continuous Processes," *Mathematical Finance*, 4, 343–348.

———, 1994b, "A General Version of the Fundamental Theorem of Asset Pricing," *Mathematical Annals*, 300, 463–520.

———, 1995, "The Existence of Absolutely Continuous Local Martingale Measures," *Annals of Applied Probability*, 5, 926–945.

Duffie, D., and K. Singleton, 1997, "An Econometric Model of the Term Structure of Interest-Rate Swap Yields," *The Journal of Finance*, 52, 1287–1321.

———, 1999, "Modelling Term Structures of Defaultable Bonds," *The Review of Financial Studies*, 12, 687–720.

Dumas, B., J. Fleming, and R. Whaley, 1998, "Implied Volatility Functions: Empirical Tests," *The Journal of Finance*, 53, 2059–2106.

Emanuel, D., 1983, "Warrant Valuation and Exercise Strategy," *Journal of Financial Economics*, 12, 211–235.

Ericsson, J., 1997, "Credit Risk in Corporate Securities and Derivatives," Ph.D. thesis, Stockholm School of Economics.

Ericsson, J., and J. Reneby, 1996, "Stock Options as Barrier Contingent Claims," working paper, Stockholm School of Economics.

——— , 1998, "A Framework for Valuing Corporate Securities," *Applied Mathematical Finance*, 5, 143–164.

——— , 2001, "The Valuation of Corporate Liabilities: Theory and Tests," working paper, Stockholm School of Economics.

Fischer, E., R. Heinkel, and J. Zechner, 1989, "Dynamic Capital Structure Choice: Theory and Tests," *The Journal of Finance*, 44, 19–40.

Fisher, L., 1984, "Discussion of "Contingent Claims Analysis of Corporate Capital Structures: An Empirical Investigation"," *The Journal of Finance*, 39, 625–627.

Galai, D., and M. Schneller, 1978, "Pricing of Warrants and the Value of the Firm," *The Journal of Finance*, 33, 1333–1342.

Geman, H., and M. Yor, 1996, "Pricing and Hedging Double–Barrier Options: A Probabilistic Approach," *Mathematical Finance*, 6, 365–378.

Geman, H., N. E. Karoui, and J. Rochet, 1995, "Changes of Numeraire, Changes of Probability Measure and Option Pricing," *Journal of Applied Probability*, 32, 443–458.

Geske, R., 1977, "The Valuation of Corporate Liabilities as Compound Options," *Journal of Financial and Quantitative Analysis*, 5, 541–552.

——— , 1979, "The Valuation of Compound Options," *Journal of Financial Economics*, 7, 63–81.

Goldstein, R., N. Ju, and H. Leland, 2001, "An EBIT–Based Model of Dynamic Capital Structure Choice," *Journal of Business*, 74, 483–512.

Hanke, M., and K. Pötzelberger, 2002, "Consistent Pricing of Warrants and Traded Options," *Review of Financial Economics*, 11, 65–79.

Harrison, J., 1985, *Brownian Motion and Stochastic Flow Systems*. Wiley, New York.

Harrison, J., and S. Pliska, 1981, "Martingales and Stochastic Integrals in the Theory of Continuous Trading," *Stochastic Processes and Applications*, 11, 215–260.

Hull, J. C., 1997, *Options, Futures and Other Derivative Securities*, 3rd edn. Prentice Hall, Upper Saddle River, N.J.

Ingersoll, J., 1977a, "A Contingent-Claims Valuation of Convertible Securities," *Journal of Financial Economics*, 4, 289–321.

———, 1977b, "An Examination of Corporate Call Policies on Convertible Securities," *The Journal of Finance*, 32, 463–477.

Itô, K., 1951, "On Stochastic Differential Equations," *Memoirs, American Mathematical Society*, 4, 1–51.

Jarrow, R., and S. Turnbull, 1995, "Pricing Derivatives on Financial Securities Subject to Credit Risk," *The Journal of Finance*, 50, 53–85.

Jarrow, R., D. Lando, and S. Turnbull, 1997, "A Markov Model for the Term Structure of Credit Risk Spreads," *The Review of Financial Studies*, 10, 481–523.

Jones, E. P., S. Mason, and E. Rosenfeld, 1984, "Contingent Claims Analysis of Corporate Capital Structures: an Empirical Investigation," *The Journal of Finance*, 39, 611–625.

Kabanov, Y., and D. Kramkov, 1994, "No-Arbitrage and Equivalent Martingale Measures: an Elementary Proof of the Harrison-Pliska Theorem," *Theory of Probability and Its Applications*, 39, 523–527.

———, 1998, "Asymptotic Arbitrage in Large Financial Markets," *Finance and Stochastics*, 2, 143–172.

Kim, J., K. Ramaswamy, and S. Sundaresan, 1993, "Does Default Risk in Coupons Affect the Valuation of Corporate Bonds? – A Contingent Claims Model," *Financial Management*, 117–131.

Klein, I., and W. Schachermayer, 1996, "Asymptotic Arbitrage in Non-Complete Large Financial Markets," *Theory of Probability and Its Applications*, 41, 927–934.

Leland, H. E., 1994, "Corporate Debt Value, Bond Covenants, and Optimal Capital Structure," *The Journal of Finance*, 49, 1213–1252.

———, 1998, "Agency Costs, Risk Management, and Capital Structure," *The Journal of Finance*, 53, 1213–1243.

Leland, H. E., and K. Toft, 1996, "Optimal Capital Structure, Endogenous Bankruptcy, and the Term Structure of Credit Spreads," *The Journal of Finance*, 51, 987–1019.

Longstaff, F., and E. Schwartz, 1995, "A Simple Approach to Valuing Risky Fixed and Floating Rate Debt," *The Journal of Finance*, 50, 789–819.

Mella-Barral, P., and W. Perraudin, 1997, "Strategic Debt Service," *The Journal of Finance*, 52, 531–556.

Merton, R. C., 1973, "Theory of Rational Option Pricing," *Bell Journal of Economics and Management Science*, 4, 141–183.

———, 1974, "On The Pricing of Corporate Debt: The Risk Structure of Interest Rates," *The Journal of Finance*, 29, 449–470.

Modigliani, F., and M. Miller, 1958, "The Cost of Capital, Corporation Finance and the Theory of Investment," *American Economic Review*, 48, 261–297.

———, 1963, "Taxes and the Cost of Capital: A Correction," *American Economic Review*, 53, 433–443.

Musiela, M., and M. Rutkowski, 1997, *Martingale Methods in Financial Modelling*. Springer, Berlin.

Neftci, S., 2000, *An Introduction to the Mathematics of Financial Derivatives*, 2nd edn. Academic Press, San Diego.

Rendleman, R., and B. Bartter, 1979, "Two-State Option Pricing," *The Journal of Finance*, 34, 1093–1110.

Reneby, J., 1998, "Pricing Corporate Debt," Ph.D. thesis, Stockholm School of Economics.

Rich, D., 1994, "The Mathematical Foundations of Barrier Option-Pricing Theory," in *Advances in Futures and Options Research: Volume 7*, ed. by D. Chance, and R. Trippi. JAI Press, Greenwich, Conn., pp. 267–311.

Rogers, L., 1994, "Equivalent Martingale Measures and No-Arbitrage," *Stochastics and Stochastics Reports*, 51, 41–49.

Rubinstein, M., 1985, "Nonparametric Tests of Alternative Option Pricing Models Using All Reported Trades and Quotes on the 30 Most Active CBOE Option Classes from August 23, 1976 through August 31, 1978," *The Journal of Finance*, 40, 455–480.

Schachermayer, W., 1992, "A Hilbert Space Proof of The Fundamental Theorem of Asset Pricing in Finite Discrete Time," *Insurance: Mathematics and Economics*, 11, 1–9.

Schweizer, M., 1992, "Martingale Densities for General Asset Prices," *Journal of Mathematical Economics*, 21, 363–378.

Sharpe, W., 1978, *Investments.* Prentice-Hall, Englewood Cliffs, N.J.

Spatt, C., and F. Sterbenz, 1988, "Warrant Exercise, Dividends, and Reinvestment Policy," *Journal of Finance*, 43, 493–506.

Taqqu, M., and W. Willinger, 1987, "The Analysis of Finite Security Markets Using Martingales," *Advances in Applied Probability*, 19, 1–25.

Toft, K. B., and B. Prucyk, 1997, "Options on Leveraged Equity: Theory and Empirical Tests," *The Journal of Finance*, 52, 1151–1180.

Vasicek, O., 1977, "An Equilibrium Characterization of the Term Structure," *Journal of Financial Economics*, 5, 177–188.

Zhang, P., 1998, *Exotic Options.* World Scientific, Singapore.

Springer-Verlag
and the Environment

WE AT SPRINGER-VERLAG FIRMLY BELIEVE THAT AN international science publisher has a special obligation to the environment, and our corporate policies consistently reflect this conviction.

WE ALSO EXPECT OUR BUSINESS PARTNERS – PRINTERS, paper mills, packaging manufacturers, etc. – to commit themselves to using environmentally friendly materials and production processes.

THE PAPER IN THIS BOOK IS MADE FROM NO-CHLORINE pulp and is acid free, in conformance with international standards for paper permanency.